Dimensional Quality Management
in 12 Basic Steps

DIMENSIONAL QUALITY MANAGEMENT
IN 12 BASIC STEPS

By K.E. Rosink

First Edition, 2009
ISBN #978-0-578-03591-8

Dimensional Quality Management in 12 Basic Steps

TABLE OF CONTENTS

PREFACE

In this book, we will detail the reasons behind the 12 Essential Steps for Quality Control, identify key focus points within each particular step and provide examples of how to create and document necessary information for each step. Additionally, we will show examples of current methods used in various production environments along with the pros and cons of these methods and how they relate to the overall goal of Quality Control.

While most of the examples will focus on the automotive industry, all of the procedures mentioned can be adjusted to fit any product industry where quality control is desired.

Measurement systems shown are for example purposes only. No specific systems or manufacturers are recommended or implied as recommended in this document.

All parts and assemblies shown are for example purposes only. Some parts may have and/or have been altered to support the discussion in this document.

Finally, we will make suggestions as to when these steps should be performed, what resources are needed in the upfront planning of these steps, key background training for complete team understanding, resources recommended for the implementation of the steps during a product's life cycle and who should be involved from inception to completion.

INTRODUCTION

One major step needed for Quality Control to succeed is an entire buy-in from all parties involved in designing, processing, and production of the product. But, most importantly, the management team. The management team not only needs to show support and guidance toward quality, but needs to have in place quality incentives that put quality above quantity, where there is a complete understanding that while shipping volume is important, shipping volume without quality is expensive. Not only in the cost to repair, but also in the loss of customer satisfaction. Management speeches only go so far, if people are not graded on the quality of product leaving their station/area they tend to not focus on the quality produced.

Additionally, quality control is a pro-active process not reactive. Quality control when in place and functioning to it's fullest capabilities generates a quality product at each station or operation of the entire product processing cycle. Furthermore, quality control measurement systems are used to indicate trends of the build and alert resources (via line stoppages or paging systems, etc.) when an operation starts trending toward tolerance limits, thus allowing issues to be corrected before the end customer or next operation sees an unacceptable product.

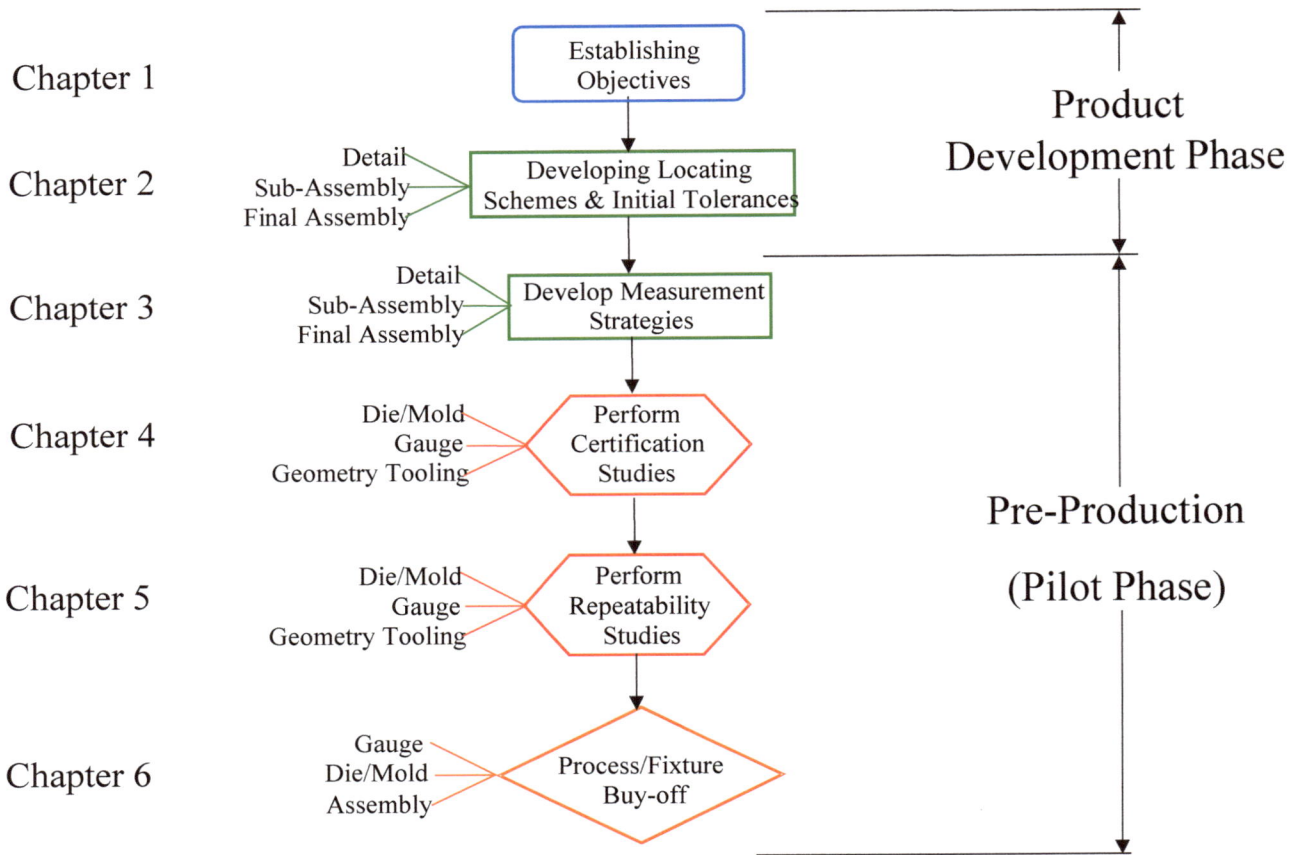

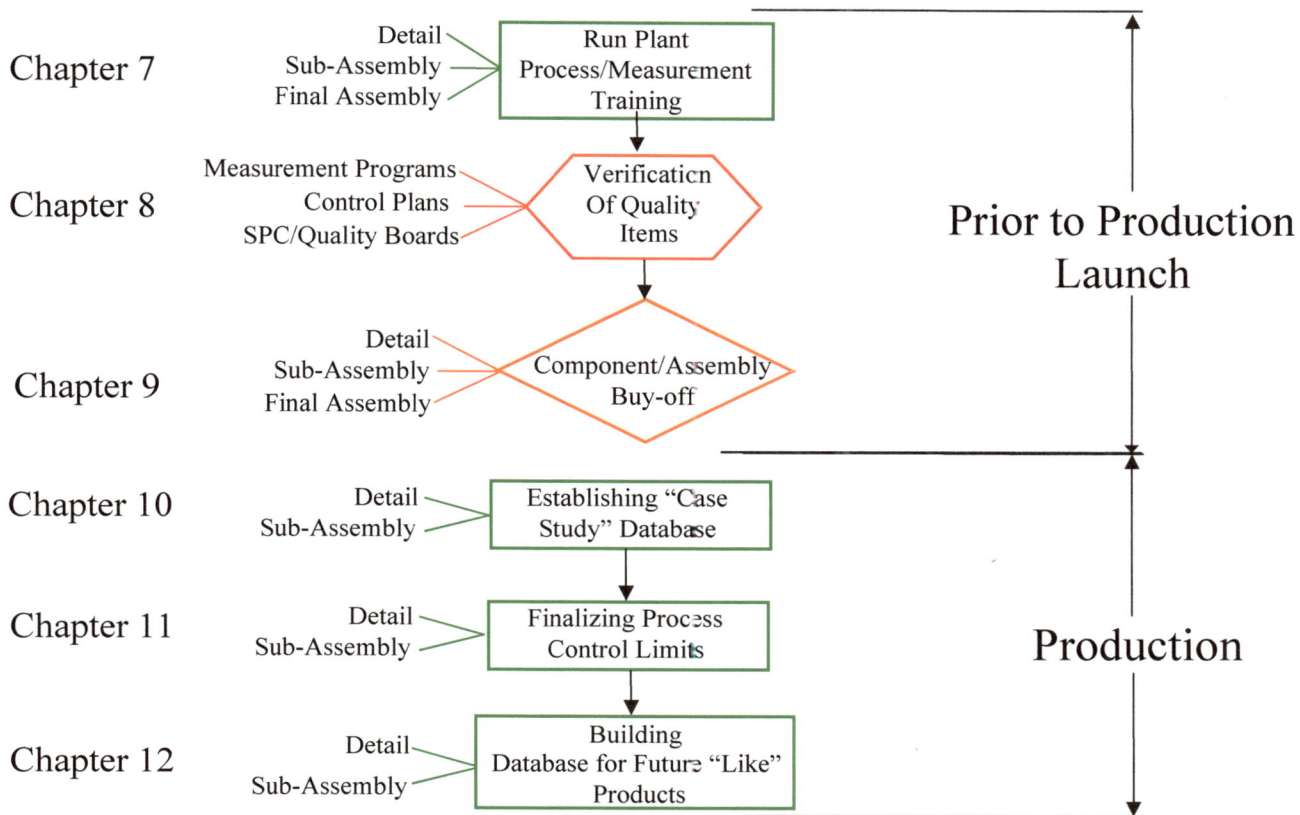

Chapter 7 — Detail, Sub-Assembly, Final Assembly → Run Plant Process/Measurement Training

Chapter 8 — Measurement Programs, Control Plans, SPC/Quality Boards → Verification Of Quality Items

Chapter 9 — Detail, Sub-Assembly, Final Assembly → Component/Assembly Buy-off

Prior to Production Launch

Chapter 10 — Detail, Sub-Assembly → Establishing "Case Study" Database

Chapter 11 — Detail, Sub-Assembly → Finalizing Process Control Limits

Chapter 12 — Detail, Sub-Assembly → Building Database for Future "Like" Products

Production

Product Development Phase

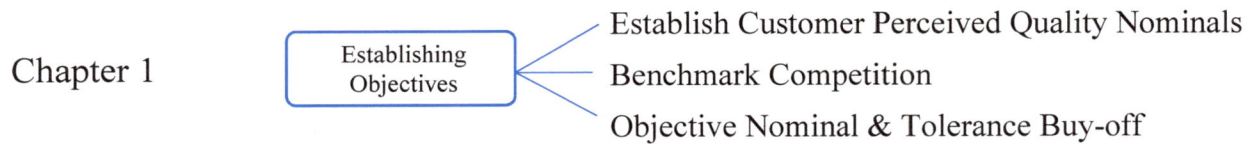

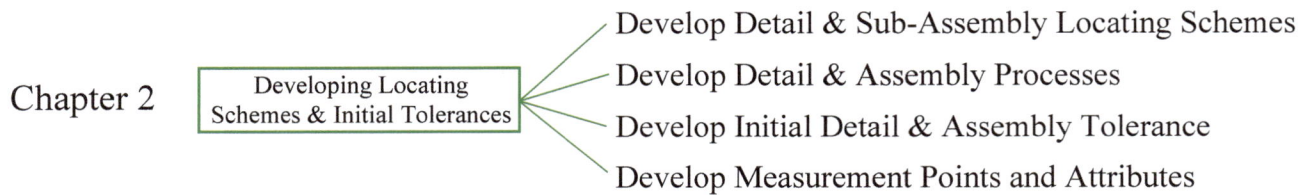

Chapter 1   Establishing Objectives

- Establish Customer Perceived Quality Nominals
- Benchmark Competition
- Objective Nominal & Tolerance Buy-off

Chapter 2   Developing Locating Schemes & Initial Tolerances

- Develop Detail & Sub-Assembly Locating Schemes
- Develop Detail & Assembly Processes
- Develop Initial Detail & Assembly Tolerance
- Develop Measurement Points and Attributes

Chapter 1 – Establishing Objectives

There are two types of objectives, dimens onal and functional. Functional objectives include door closing efforts, window roll-up speed, etc. Dimensional objectives include customer fit and finish expectations on a complete automobile, seal gap or distance on a refrigerator, to the machine finish of a surface on a VCR tape head. Additionally, other items, not fitting in the functional or dimensional category of the end customer expectations, but equally important to the final goals are zone control items, such as monitoring the individual steps outlined in a complete process.

From a dimensional management or quality control perspective, anything and everything that can influence the overall objective, should be viewed as an individual goal or target towards the overall goal.

Quality control starts with the understanding and establishment of objectives – what is the end result that will determine an acceptable versus unacceptable product. Additionally, it is important to view the objective from multiple perspectives:

- End customer perception
- Production run inspector(s)
- Processing
- Product Design
- Process Design

While each of these perspectives are important, a hierarchy for these perspectives should be established (starting with the end customer as the most important) so the entire team understands and agrees on the rankings. The team considering this should consist of the following:

- Product Design
- Assembly Process
- Stamping/Molding Process
- Dimensional Management
- Customer Audit Group
- Production run source quality department(s)

Once this is established, each team member, singularly and collectively, can start focusing on their piece of the pie completed. Numerous items need to be considered at this point:
- What does the end customer consider an acceptable product?
- What processing limitations exist?
- What methods for evaluating the level of acceptability exist?
- What design constraints exist?

In tackling these issues, we need to address these questions collectively and separately:

1.)   What does the end customer consider an acceptable product?
The answer to this question may be achieved in numerous ways including; performing benchmarking studies on competitive products or by holding a "clinic" for perspective customers to view and critique your product at the prototype stage or by creating a questionnaire for potential customers to answer. Please keep in mind that in the latter two suggestions, you need to get enough detail in the data to assist your development without creating a situation that alienates your perspective customer when answering it.

When performing benchmark studies, it is important to identify critical characteristics on the product that you and your end customer will use to signify the differences between an acceptable and unacceptable product (see figure 1). (These characteristics are explained in greater detail in chapter 4.) Additionally, benchmark studies should be performed on a minimum sample size of 30 units of the same make – this generates statistical data with a greater degree of reliability and predictability for the entire population. Benchmark studies should also utilize the same measurement devices, where possible, that will be used when verifying your own product. This helps to understand various ways an attribute can be measured using a single device and the relationship of that device from product to product.

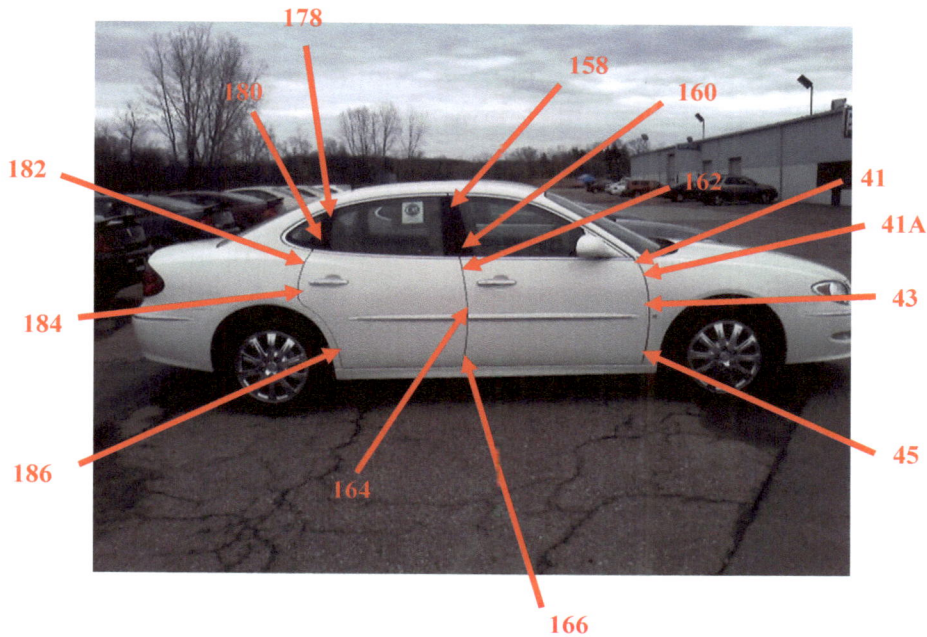

Right Side Doors Gap   (All Dimensions in mm)

				Vehicle Type #1		Vehicle Type #2		Vehicle Type #3	
POINT	FEATURE DESCRIPTION	LOCATION	CHARACTERISTIC	MEAN	± 3 SIGMA	MEAN	± 3 SIGMA	MEAN	± 3 SIGMA
41	Right Front Door to Hood/Fender	Top	Gap	4.37	± 2.42	5.49	± 3.11	5.44	± 0.78
41A	Right Front Door to Hood/Fender	Bottom	Gap	4.42	± 2.27	5.36	± 2.38	5.50	± 0.70
43	Right Front Door to Hood/Fender	Middle	Gap	5.03	± 2.10	5.69	± 1.88	5.69	± 0.98
45	Right Front Door to Hood/Fender	Bottom	Gap	4.43	± 1.91	5.09	± 2.50	5.37	± 1.12
P7	Right Front Door to Fender	Point 41-43	Parallelism	-0.65	± 2.72	-0.21	± 3.24	-0.25	± 1.26
P8	Right Front Door to Fender	Point 43-45	Parallelism	0.60	± 2.55	0.60	± 2.76	0.33	± 0.88
89	Right Front Door to Quarter Panel	Top of Window	Gap	-	-	5.21	± 2.67	-	-
91	Right Front Door to Quarter Panel	Middle of Window	Gap	-	-	5.24	± 2.22	-	-
93	Right Front Door to Quarter Panel	Top of Door Panel	Gap	-	-	4.75	± 1.52	-	-
95	Right Front Door to Quarter Panel	Middle of Door Panel	Gap	-	-	5.35	± 1.18	-	-
97	Right Front Door to Quarter Panel	Bottom of Door Panel	Gap	-	-	5.20	± 1.18	-	-
P9	Right Front Door to Quarter Panel	Point 89-91	Parallelism	-	-	-0.03	± 3.47	-	-
P10	Right Front Door to Quarter Panel	Point 91-93	Parallelism	-	-	0.49	± 2.14	-	-
P11	Right Front Door to Quarter Panel	Point 93-95	Parallelism	-	-	-0.60	± 1.41	-	-
P12	Right Front Door to Quarter Panel	Point 95-97	Parallelism	-	-	0.15	± 1.60	-	-
158	Right Front Door to Rear Door	Top of Window	Gap	4.88	± 3.90	-	-	5.23	± 1.11
160	Right Front Door to Rear Door	Middle of Window	Gap	5.29	± 2.28	-	-	5.52	± 0.94
162	Right Front Door to Rear Door	Top of Door Panel	Gap	5.25	± 1.87	-	-	5.96	± 0.66
164	Right Front Door to Rear Door	Middle of Door Panel	Gap	5.24	± 1.80	-	-	5.71	± 0.79
166	Right Front Door to Rear Door	Bottom of Door Panel	Gap	5.13	± 2.05	-	-	5.72	± 1.15
P13	Right Front Door to Rear Door	Point 158-160	Parallelism	-0.41	± 3.27	-	-	-0.29	± 1.05
P14	Right Front Door to Rear Door	Point 160-162	Parallelism	0.04	± 2.46	-	-	-0.43	± 0.99
P15	Right Front Door to Rear Door	Point 162-164	Parallelism	0.01	± 2.16	-	-	0.24	± 0.71
P16	Right Front Door to Rear Door	Point 164-166	Parallelism	0.11	± 2.14	-	-	0.00	± 0.92
178	Right Rear Door to Quarter Panel	Top of Window	Gap	5.74	± 3.24	-	-	-	-
180	Right Rear Door to Quarter Panel	Middle of Window	Gap	5.06	± 3.69	-	-	-	-
182	Right Rear Door to Quarter Panel	Top of Door Panel	Gap	4.40	± 1.90	-	-	-	-
184	Right Rear Door to Quarter Panel	Middle of Door Panel	Gap	5.33	± 1.77	-	-	-	-
186	Right Rear Door to Quarter Panel	Bottom of Door Panel	Gap	5.35	± 2.12	-	-	-	-
P17	Right Rear Door to Quarter Panel	Point 178-180	Parallelism	0.68	± 3.20	-	-	-	-
P18	Right Rear Door to Quarter Panel	Point 180-182	Parallelism	0.67	± 3.19	-	-	-	-
P19	Right Rear Door to Quarter Panel	Point 182-184	Parallelism	-0.93	± 2.24	-	-	-	-
P20	Right Rear Door to Quarter Panel	Point 184-186	Parallelism	-0.03	± 2.18	-	-	-	-

Figure 1 – Benchmark Study example

In the figure above, measurements are made on the critical characteristics noted on the picture for all vehicles measured. Data shown is the result of a 30-sample per vehicle type study. In this example, 3 different types of vehicles were measured.

One additional note on benchmark studies --- data retrieved should be with the understanding that it is representative of what your competitor is currently delivering to its end customer and by no means is an indication of what your competitor produces out of a given process. What I mean is that the end result you are measuring does not indicate one particular process your competition used to produce this condition but possibly many. Some of these you are aware of (by benchmark trips to various competitor facilities in key processing areas) and some you aren't (out of sequence processes that benchmarking trips often don't uncover).

2.)   What processing limitations exist?

Processing limitations, that is the limitations surrounding the entire process of making the product itself, can be as simple as monies available for tooling, designated floor space available, or more intense issues such as cycle time constraints and utilizing carry-over processes for new products. While these are only a few of the many possibilities it is important to identify all the things that are available to establish the process for the product.

Another key element that falls into the processing issue, but is not necessarily a limitation is the opportunity for process benchmarking. Like the measurement studies mentioned before, this is an opportunity to evaluate a competitor's process that achieves a similar end result as you desire. It is important to distinguish and evaluate all the processes in a detailed manner to understand what parts of the process are automated, manual, fixtured, hand-fit, how many stations, where the process locators are at each processing operation, torque sequences, clamp sequences, etc. if this detail can be observed.

3. What methods for evaluating the level of acceptability exist?
Here we are talking about the measurement resources available. Multiple answers to this question are possible based on the time frame of the product cycle. Initially, at the prototype stage, and at initial production product buyoff more measurement may be required to establish projected process capability. At this stage, variable data is essential to determine initial process capability. Also at this stage, data is usually generated from a single product production run (see figure 2), where possible, however, it is advantageous to gather additional data from additional product production runs (also see figure 2). This gives a better understanding of the overall process capability not just from the tools creating the product themselves. What this means is, for example, a car door inner panel measured from a single die run will show the process capability, mean and variation of that particular die set. Additional factors including, multiple die sets, die wear, etc. including the system automation between each die and prior to loading the panels into the shipping rack will not be witnessed in the variation of a single die run. Over multiple die runs, these added processes, all within the normal production process cycle, will show additional variation. Measurement data from multiple production cycles (or runs) will help illustrate the long term expected process capability and provide better upfront knowledge when determining tolerances for a complete product including all of its processes.

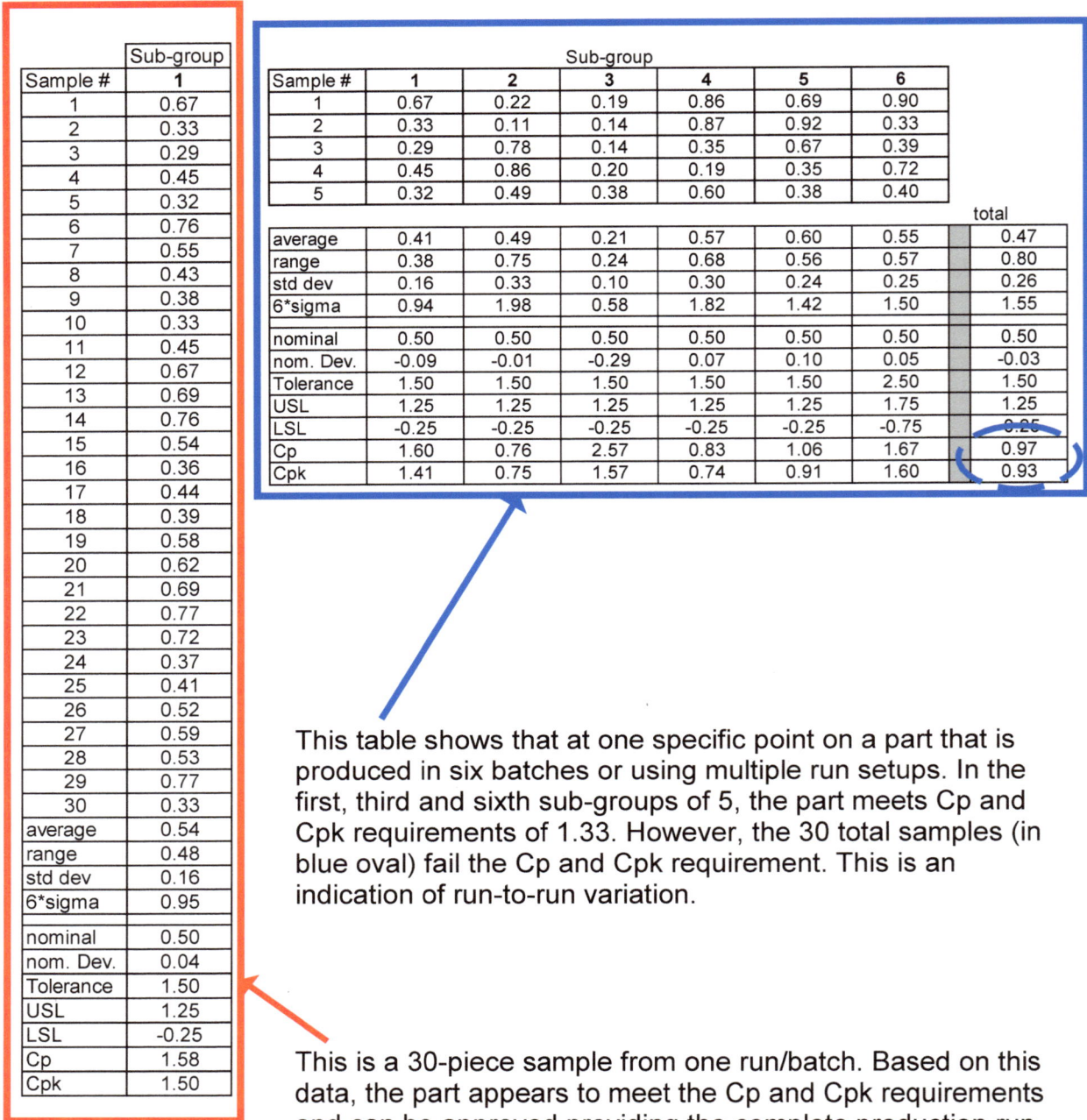

Sample #	Sub-group 1
1	0.67
2	0.33
3	0.29
4	0.45
5	0.32
6	0.76
7	0.55
8	0.43
9	0.38
10	0.33
11	0.45
12	0.67
13	0.69
14	0.76
15	0.54
16	0.36
17	0.44
18	0.39
19	0.58
20	0.62
21	0.69
22	0.77
23	0.72
24	0.37
25	0.41
26	0.52
27	0.59
28	0.53
29	0.77
30	0.33
average	0.54
range	0.48
std dev	0.16
6*sigma	0.95
nominal	0.50
nom. Dev.	0.04
Tolerance	1.50
USL	1.25
LSL	-0.25
Cp	1.58
Cpk	1.50

Sample #	Sub-group 1	2	3	4	5	6		total
1	0.67	0.22	0.19	0.86	0.69	0.90		
2	0.33	0.11	0.14	0.87	0.92	0.33		
3	0.29	0.78	0.14	0.35	0.67	0.39		
4	0.45	0.86	0.20	0.19	0.35	0.72		
5	0.32	0.49	0.38	0.60	0.38	0.40		
average	0.41	0.49	0.21	0.57	0.60	0.55		0.47
range	0.38	0.75	0.24	0.68	0.56	0.57		0.80
std dev	0.16	0.33	0.10	0.30	0.24	0.25		0.26
6*sigma	0.94	1.98	0.58	1.82	1.42	1.50		1.55
nominal	0.50	0.50	0.50	0.50	0.50	0.50		0.50
nom. Dev.	-0.09	-0.01	-0.29	0.07	0.10	0.05		-0.03
Tolerance	1.50	1.50	1.50	1.50	1.50	2.50		1.50
USL	1.25	1.25	1.25	1.25	1.25	1.75		1.25
LSL	-0.25	-0.25	-0.25	-0.25	-0.25	-0.75		-0.25
Cp	1.60	0.76	2.57	0.83	1.06	1.67		0.97
Cpk	1.41	0.75	1.57	0.74	0.91	1.60		0.93

This table shows that at one specific point on a part that is produced in six batches or using multiple run setups. In the first, third and sixth sub-groups of 5, the part meets Cp and Cpk requirements of 1.33. However, the 30 total samples (in blue oval) fail the Cp and Cpk requirement. This is an indication of run-to-run variation.

This is a 30-piece sample from one run/batch. Based on this data, the part appears to meet the Cp and Cpk requirements and can be approved providing the complete production run of the part is done in one batch. If this part is manufactured in more than one run, however, this data may not indicate what parts from the other runs or batches may be produced at.

Figure 2

In both cases shown above, neither data gathering technique is incorrect provided that the data gathered is:
- based off the entire batch of parts if from one run of parts
- based off multiple batches if parts are produced in multiple runs.

4. What design constraints exist?
When referring to "design constraints" it is important to know and understand what limitations are inherent in the product design itself. Is the product partially a "carryover" product, meaning most of the design from a product already produced with some changes. Or is the product entirely new?

A carryover product also carries the "sins of the past" with it. The reason it is not changing maybe a limitation in resources issue, however, the fact remains that since it is a carryover product, that product also carries the previous tolerancing, locating schemes and processing from the product similarily produced. What this means is that instead of "re-inventing the wheel", tweak it where known unacceptable quality levels exist by improving the design and/or the process to meet the new quality objectives.

Establishing Objectives – Example 1

In the design of an automotive front door to rear door fit, the design office recommends a 0.0mm flush condition (see figure 5). While this nominal seems acceptable, benchmark studies on competitive vehicles (see figure x) are performed to determine current customer expectation levels and to better understand what the customer might expect when this vehicle is launched or in production 2-3 years from now. Upon gathering the benchmarking data, a "design buck" (see figures 3 & 4) may also be created – this will be used to show "high" and "low" tolerance specifications that will be assigned to the objective nominal. Another method of showing high and low tolerance specifications is to alter the CAD model by moving one part with respect to the other – while this method works, it does not allow people to view the area in question at the same sight height and locations that a customer would. A team consisting of :

- Stamping production plant quality manager
- Assembly plant quality manager
- Product design
- Design Studio
- Advanced Processing engineering
- Dimensional Management

gathers and reviews the data and the buck. If all parties agree that the buck represents end customer acceptance criteria, the objective is documented and bought off by the entire team (see figure 6). If issues arise, multiple tasks are generated and new mock-ups are created on the buck or CAD model until a full team consensus is achieved.

Once achieved, signatures are collected and logged into a database for future reference. Tolerances are then rolled down to the respective assemblies to finally the details themselves.

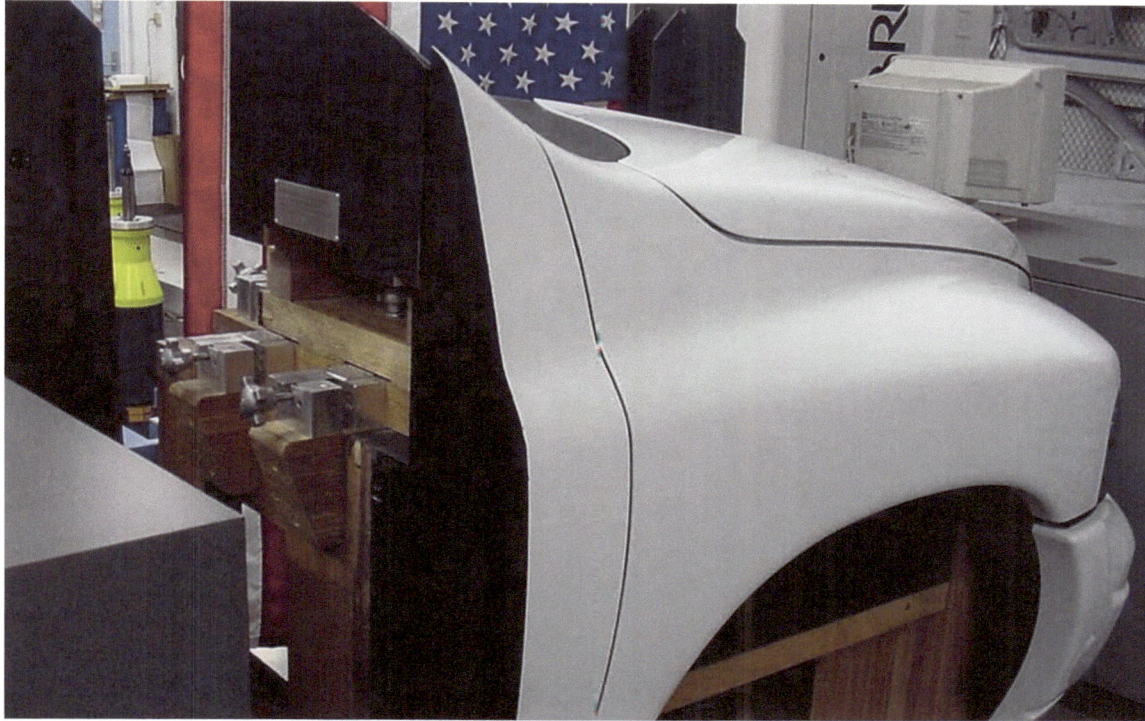

Figure 3 – Design Buck

Figure x, shows a design buck that compares hood and fender flush and gap fit. Here the hood and fender can be adjusted individually to the desired tolerance specifications for review. An important note for this buck is that the hood height is the same as if it was a completed vehicle on wheels-this gives the team the proper viewing perspective that a customer would expect to see. A better view of the adjustment devices is found in figure 4 (next page).

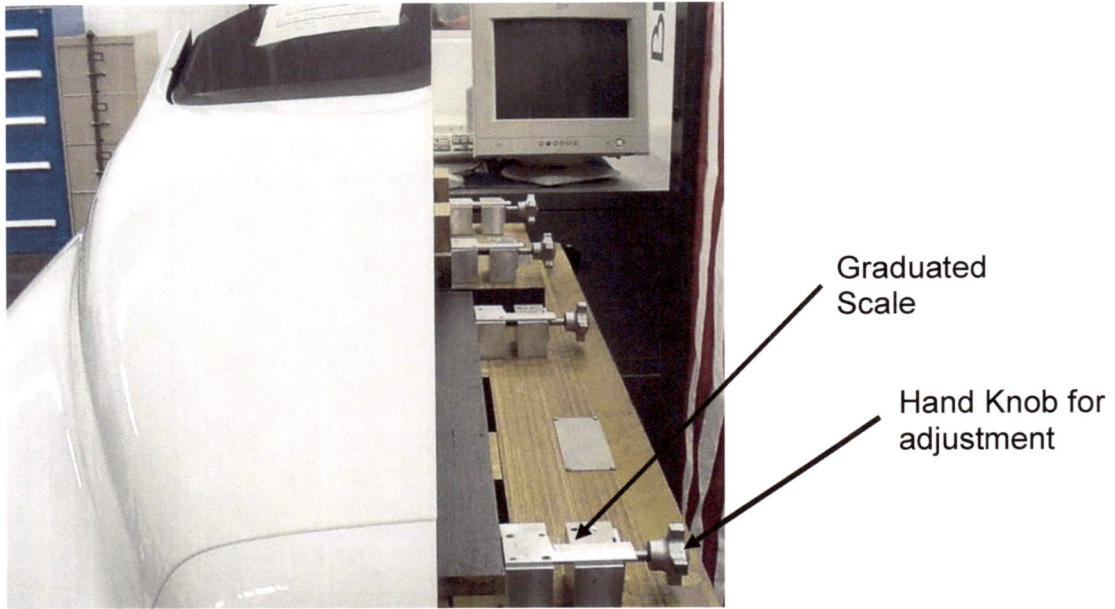

Figure 4 – Design Buck Adjustment Features

Here, hand knobs can be turned in either direction to move the hood inboard or outboard of the fender. A graduated scale (in millimeters) indicates the amount of movement that the hood has made in this direction. However, it is important to note that while the graduated scale may indicate a 3.00mm move, the actual flush measurement should be made using the prescribed measurement device from the objective callout. This is necessary as the production measurement device may align differently than the graduated scale, thus giving a different result.

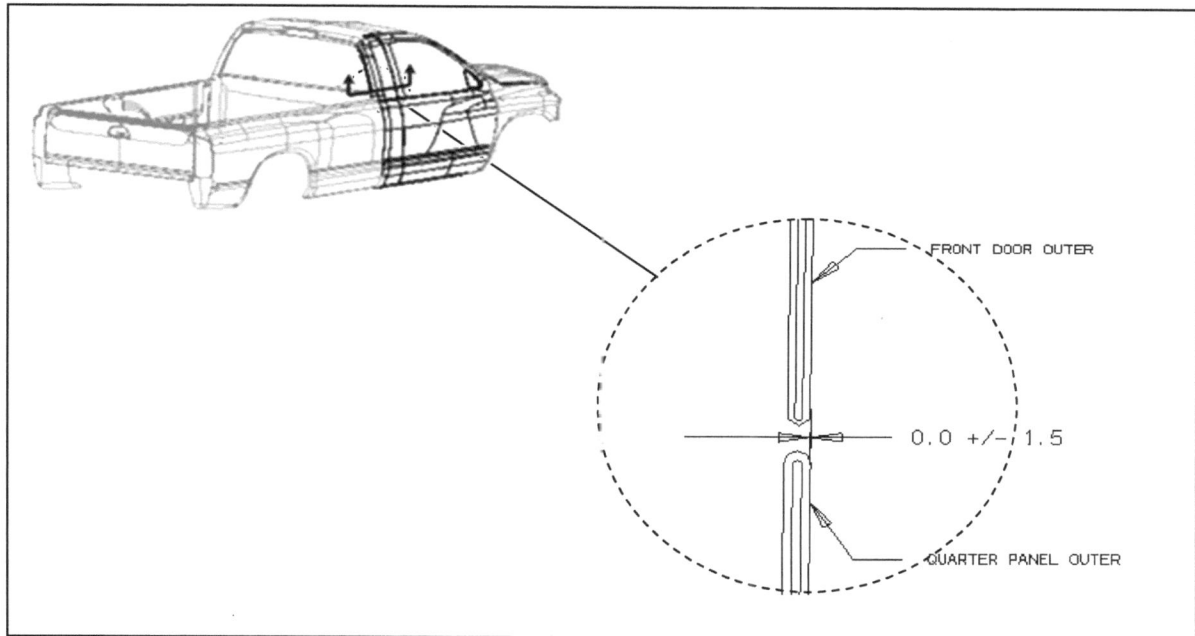

Figure 5 – Objective Documentation

Here the team has determined that the flush or surface to surface measurement between the front door and the rear door should be even or a nominal of 0.0mm. Once this is established, the doors on a buck or CAD (Computer Aided Design) model can be adjusted, based on variation data gathered from benchmarking studies, for the team to determine the maximum and minimum one door can be (overflush) to the other. Again, with a team consensus on what the end customer would still perceive as a quality product, the tolerance for the objective is established – in this instance +/-1.5mm.

At this point, the objective can be submitted for signatures to all parties, this ensures that everyone, from initial product design, through process design and finally the product run source(s) understand and agree with the quality level goals the team will design and process to meet.

```
┌─────────────────────────────────────────────────────────────────────────┐
│BUILD OBJECTIVE -                                                          │
│                                                                           │
│Number:     BIW                         Status:     Approved  4/11/1995    │
│Objective:  Rear Door To Front Door Flush                                  │
│Responsible Team:  N/A  Type:   Customer                                   │
│Engineering Area:  Body In White        Risk Analysis:                     │
│Approver:                               Consensus Date:                    │
│Affected                                                                   │
│Departments:                            Release Level:                     │
│Specification Description:    Flush                                         │
│                                                                           │
│Nominal:     0.0  Tolerance:    +/- 1.5                                     │
│                                                                           │
│                                                                           │
│THE SIGNATURES BELOW ACKNOWLEDGE THAT THE TOLERANCES SHOWN REPRESENT THE    │
│MAXIMUM VARIATION ACCEPTABLE TO MEET OR EXCEED THE CUSTOMER'S PERCEPTION OF │
│QUALITY.                                                                   │
│                                                                           │
│CRITICAL BUY OFF TEAM MEMBERS ARE SHOWN IN RED                             │
│                                                                           │
│Buy Off Team:                                                              │
│Name                    Department            Status      Date             │
│Joe Smith               Product Engineering   Accept      03/13/1995       │
│Ray Lone                Dimensional Control   Accept      03/07/1995       │
│Hans Janski             Assembly Plant #1     Accept      03/07/1995       │
│Bill Base               Assembly Plant Quality Accept     03/07/1995       │
│Jon Carper              Stamping Plant Quality Accept     03/15/1995       │
│Leon Zenolwicz          Stamping Process Eng'r Accept     03/21/1995       │
│Mike Donner             Assembly Process Eng'r Accept     03/14/1995       │
│                                                                           │
│                                                                           │
│Created:       08/31/1994          Last Modified:     12/11/1995           │
└─────────────────────────────────────────────────────────────────────────┘
```

Figure 6 – Objective Documentation with Electronic Signoff

As shown, in figure 6, all team members have accepted the objective via an electronic sign off. This document resides in a database with access available to all parties involved in producing this product.

To recap this section, when establishing objectives the following items needs to be addressed and/or determined.

1) Set customer design nominals
2) Establish Objectives Team (consists of part design, run plant quality, process design, etc.)
3) Benchmark the competition – process methods
4) Benchmark the competition – measurement studies
5) Set customer design tolerances
6) Establish process parameters (carryover, all new, etc.)
7) Establish part parameters (carryover, all new, etc.)
8) Achieve complete team objective buyoff

Chapter 2 – Developing Locating (Datum) Schemes

Locating (Datum) schemes are developed for two main reasons. First, is to hold the part for measurement and second, to hold the part for processing. In either of these two situations, it is important to remember not to over-constrain the part (i.e. do not apply more locators than needed to properly support the part consistently). The basis of these two reasons is the need to maintain a similar relationship; the locating scheme used to measure the part should be the same as the locating scheme used to process the part. The locating scheme must be the best choice to serve as the origin for measuring the part and must be capable of being used when adjoining this part to another in a fastening operation.

The ultimate desire is to achieve a correlation between the variation measured in the detail part, the variation contributed by this part to the assembly, and the variation it contributes to the complete product.

Having stated this, there are still a few key considerations that effect whether a common locating scheme can be maintained from measurement through processing. One major exception is floor line or gravity effects. In some cases, a part may be measured in a different position with respect to the plant floor versus the location of the part during processing. Another exception is the processing or tooling itself.

Locators needed to measure the part may not be feasible due to clearance considerations for weld guns, nut runners, etc. In this case, it is important to note that if a locator is absolutely necessary in the measurement stage to pass repeatability criteria, then one can expect increased variation when a locator in the same area cannot be achieved in the process. The best locating scheme is one that can be the same for measurement and process regardless of floor line.

Three main considerations in locating scheme selection involve geometry, function, and process. The selected scheme will be a compromise which best accommodates all of these considerations.

Initial thoughts prior to developing a locating scheme are to remember the 3-2-1 rules. That is that the locating scheme needs to locate on three mutually orthogonal planes, controlling six degrees of freedom. Ideal geometry includes selecting the largest triangle on the largest plane for the primary direction (3 points), selecting the longest line contained on one of the remaining planes for the secondary direction (2 points) and selecting a point on the last plant for the tertiary direction (1 point). Some parts, based on geometry and rigidity, may require more than 3 points to establish a plane. While some other parts may need more than 2 points to establish the line, primary reasons are often due to gravity sag and part rigidity issues. In these cases it may be necessary to utilize FEM (Finite Element Modeling) to assist in picking the best locations for additional locating points.

Below is an example of the use of FEM for datums. Figure 7 portrays the original creation of datums or locators. These datums are used in the finite element model to distinguish where the constraints are and in what direction. After the model is "meshed", a force of 1.5 Gs or 1.5 times the force of gravity is applied; the FEM analysis indicates "hot spots" or areas of deflection. These areas indicate where the part is most likely to sag and to what degree if the part were supported at the locators in a gauge or tooling fixture (see figure 8). Any deflection greater that 0.50mm, in this example, has required an additional support to be added in the gauge and assembly tooling process (denoted in figure 9). At this point, the FEM analysis is re-run to confirm expected results. This step reduces cost of tooling and gauge changes by verifying in the model before changing the design and build.

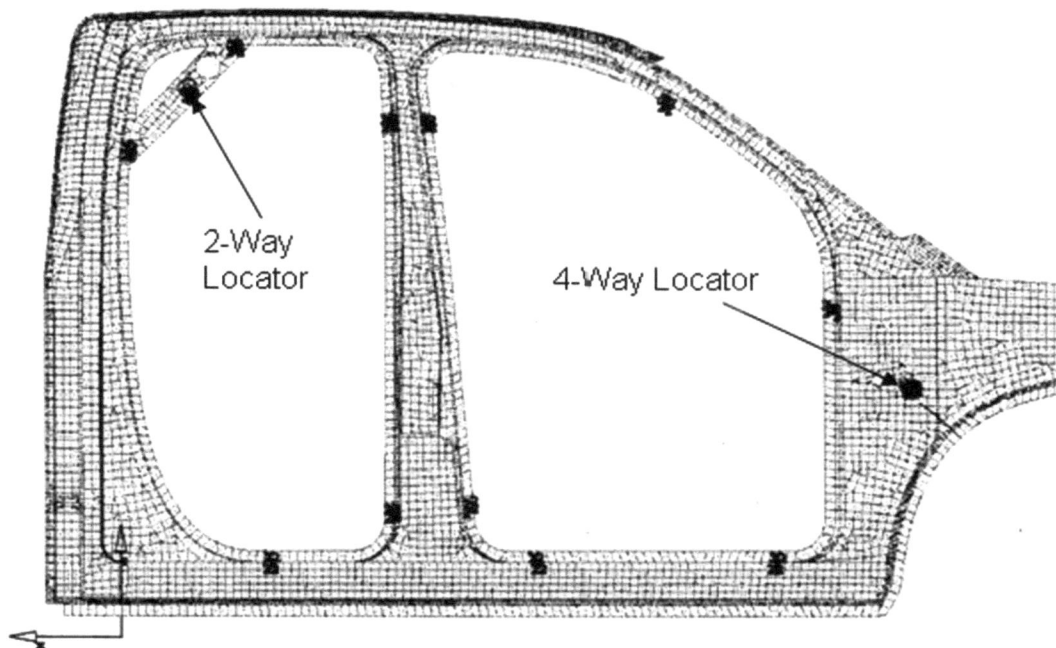

Figure 7 – Original Datum scheme

In this figure, the datums used in the FEA model are shown in black. In this model, the bodyside is 90° from body position, which is the position the part, will be processed in. Each datum, with the exception of the 2-way and 4-way locators, is constrained in 2 directions. The 4-way in constrained in 4 directions and the 2-way is constrained in 2 directions.

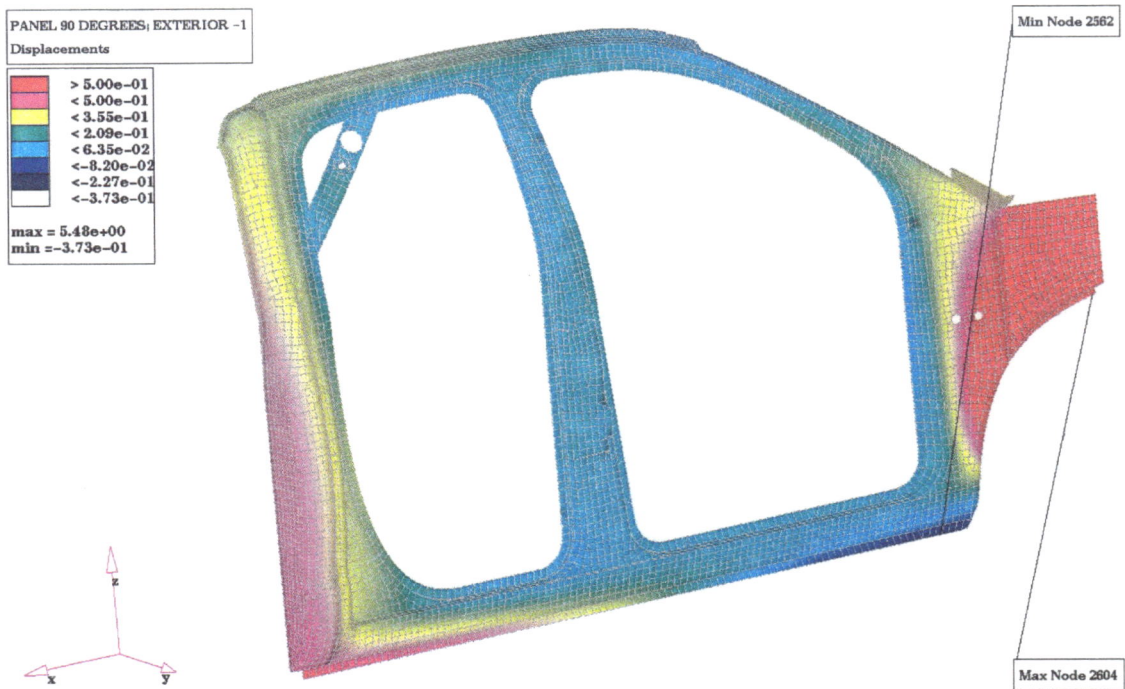

Figure 8 – Finite Element Analysis Results using original datum scheme

Here, the colors represent deflection from math model nominal when the 1.5G load is applied. The goal is to have less than 0.50mm. This goal is with the understanding that unless you create a complete "net" fixture, meaning that the entire part rests on a locator, some sag will always exist. In the figure above, it is important to note that the maximum deflection is 5.48mm (denoted by "max node 2604). The color chart at the upper left also indicates areas of "red" with greater than 0.50mm sag or deflection.

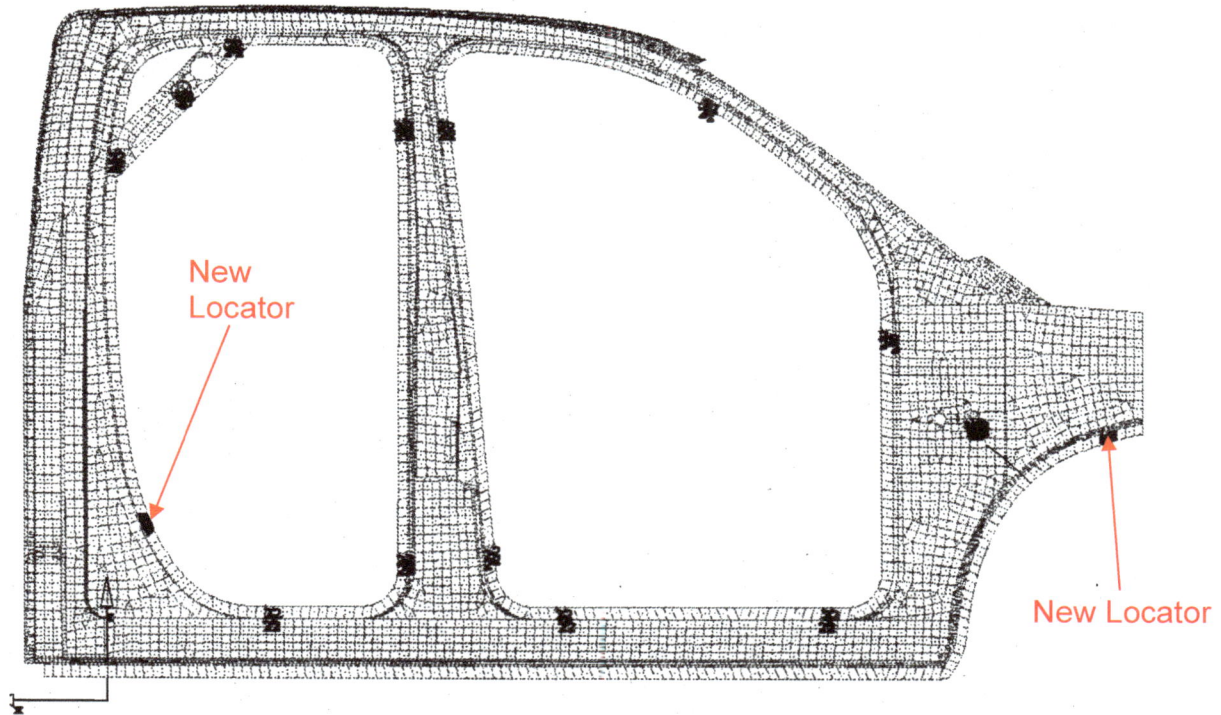

Figure 9 – Original Datums with (2) New Datums

In this spectrum, two additional datums are added for this floor line position to reduce the amount of sag. The placement of these new locators is based on best geometry for the given deflection results, functionality based on mating surfaces, and access during the assembly process.

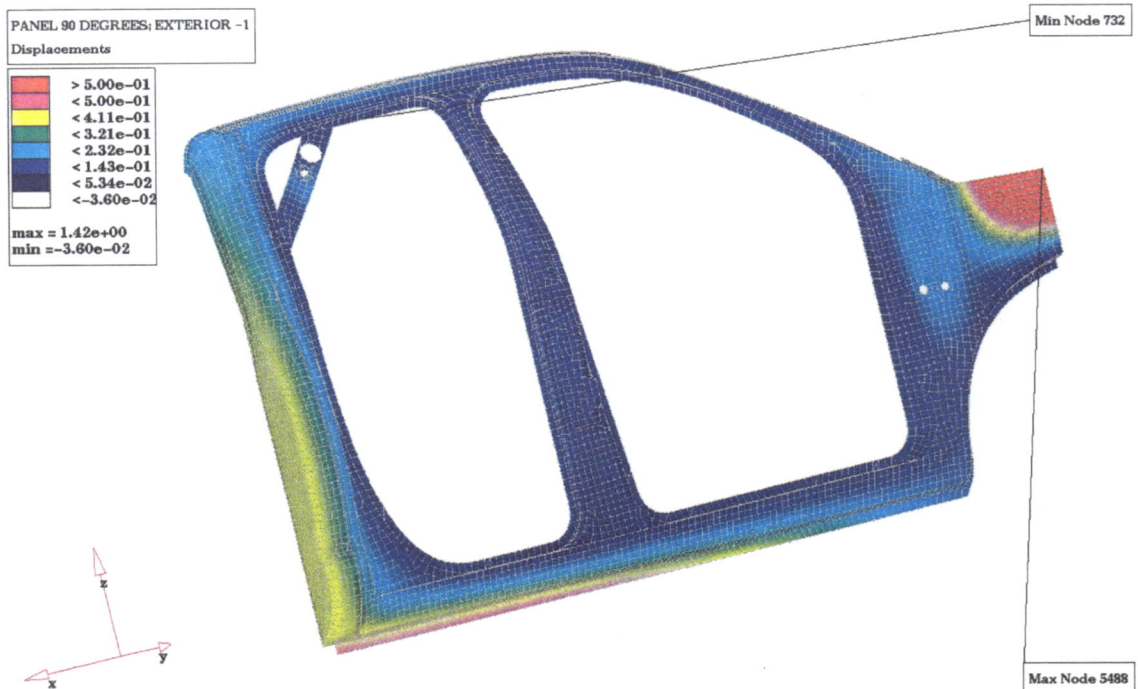

Figure 10 – FEM results with new datums

In figure 10 above, the "hot" or most deflected areas from the previous FEA model have been reduced to the "max node 5488" which has a deflection of 1.42mm. Since the surface that is non-rigid, can be welded later in the assembly process when it will be in body position, it was deemed acceptable in this example.

For each part or assembly, the locating scheme must be on the main body of the part. Placement of locator points on wings, fingers, extensions, and appendages is to be avoided. Additionally, a single feature chosen as a locator must not conflict with any other locators. It is important to note that whatever part you are measuring/processing the minimum possible locators needed to pass repeatability should be the goal. Over-constrained parts due to too many locators can add stress to the process and cover up issues during measurement thus hampering efforts to root cause any build issues.

It is recommended that the locators be positioned on flat surfaces and placed square to the body grid system. This step will assist the production plant in tooling adjustments (commonly referred to as "shim moves") over the production life cycle by eliminating, or reducing, secondary plane effects of an adjustment to the primary plane.

Additionally, it is suggested that the locators be placed on features of the part that are developed as early as possible in the part making process. Stated this way, primary plane locators should be selected on the surface that is created in the initial part making stage. This theory should also be true for the secondary and tertiary locators, so that they may be used as early as possible in all subsequent process operations. An example of this situation is found in figure x.

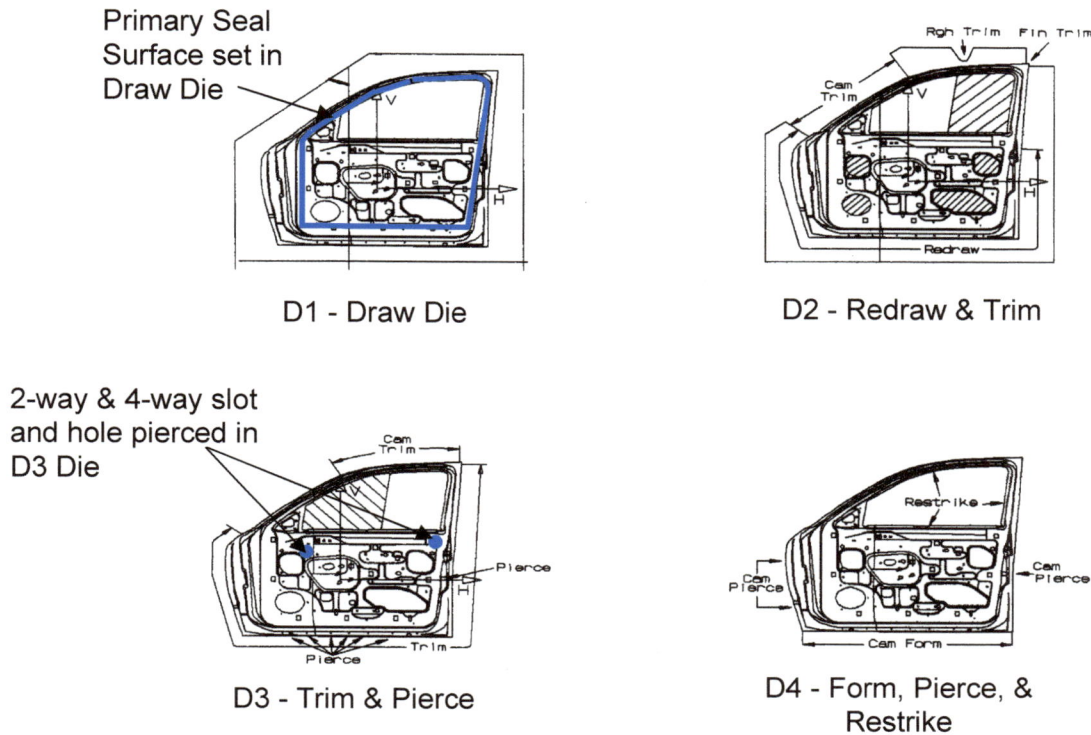

Figure 11 – Die Process of door inner panel

For a door inner panel, the primary seal surface is where the cross/car datums are located. This surface is drawn "home" in the D1 die or draw die. This also helps door build root cause issues since everything measured is relative to the seal surface and the first part operation. The 2-way and 4-way features, a slot and a hole are also set in the first pierce die, that being the D3 die (in this example). This allows these holes to be utilized for die gaging (part locators) in the D4 Die.

Surface locators for the primary plane usually are best, however, holes and slots for the secondary and tertiary plane may be more beneficial than surface locators for these directions. This is due in part, to the fact that the secondary surface relationship to the primary locating surface is hard to control across the entire line generated by the secondary locators. Thus, making the locating scheme less repeatable. An example of this comparison is depicted in figure 11.

Finally, when using hole (4 direction locator) and slot (2 direction locator) it is recommended that a MMC or Maximum Material Condition be selected. This enables the measurement device to utilize a similar pin to that in the processing tools.

Another option is the RFS or Regardless of Feature Size locator, which basically means that no matter the size of the hole or slot within the specified tolerance, the locator will be the center of the feature in the direction of the locator (see figure x below). One side note if selecting MMC is that most CMM's (Coordinate Measurement Machines), when aligning to the part itself, use a RFS algorithm so repeatability results from this device may hide part locating issues that would normally be uncovered by the "float" between a MMC pin and hole condition.

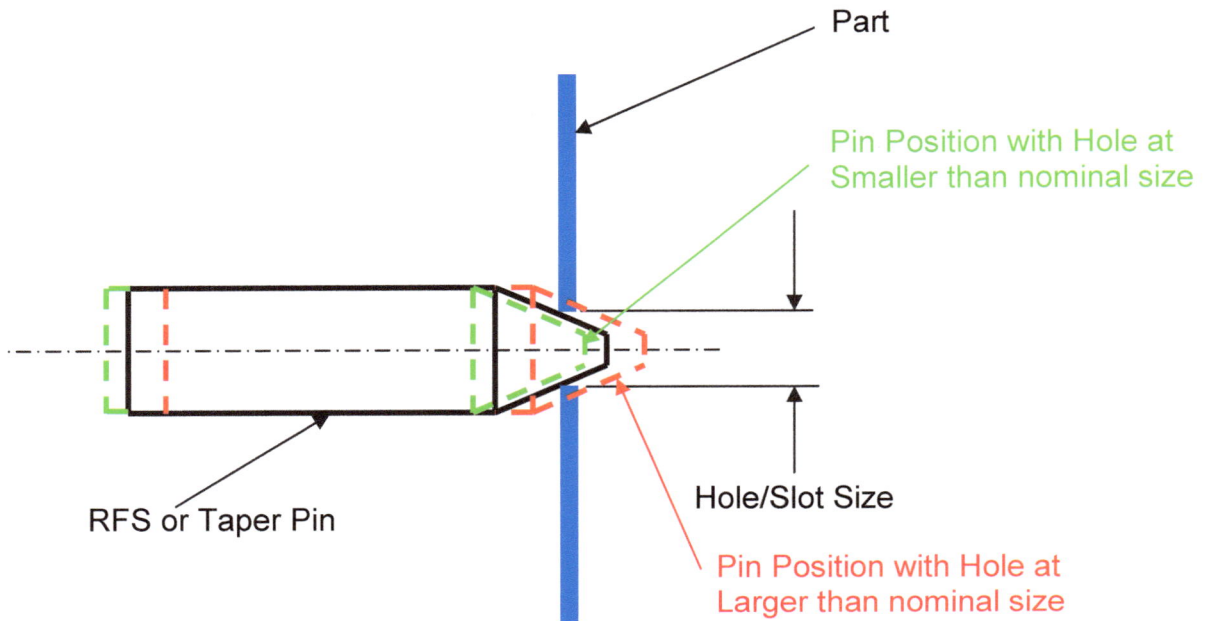

Part

Pin Position with Hole at Smaller than nominal size

RFS or Taper Pin

Hole/Slot Size

Pin Position with Hole at Larger than nominal size

Taper Pin in a RFS condition is spring loaded to allow the pin to move to the size of the hole. The axis of the pin will always align to the center of the hole/slot in the locating direction.

Figure 12 – RFS pin to hole condition

Here, if gravity loads are directly on the pin (perpendicular to the pin axis) the part may slide off the taper of the pin. In this instance, a surface locator surrounding the pin may be necessary to hold the part on the pin. Another method may be to change the floorline of the fixture to be parallel with the part surface. This process will eliminate any locating issues due to gravity effects.

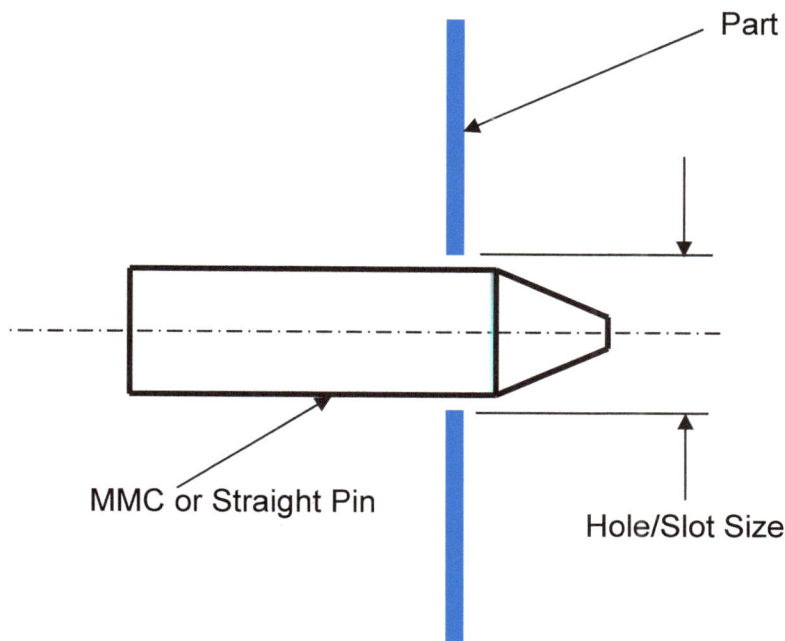

Straight Pin in a MMC condition is solid
mounted. Here, the part hole will have
"float" or slop to the pin.

Figure 13 – MMC to hole condition

If gravity loads are directly on the pin (perpendicular to the pin axis), in this
example, the part will nest on the top of the pin. Stated differently, the upper
portion of the hole or slot will rest directly on the top of the pin.

Once a locating scheme is selected and proven out, to the point where the
scheme has passed all repeatability requirements (see chapter 3 for an in-depth
discussion on repeatability), it is recommended that the locating scheme be
documented so future products can use this as a starting base.

One example of floorline or gravity effects can be witnessed on a bodyside or aperture outer detail. In a ring gage fixture the aperture is positioned in body position (see figure x). In this instance, the primary gravitational forces are on the up/down locators, the 2-way and 4-way pins. The aperture outer, however, is processed in a laydown position or 90 degrees from body position (see figure 15). Here, the gravitational forces are primarily on the cross/car locators.

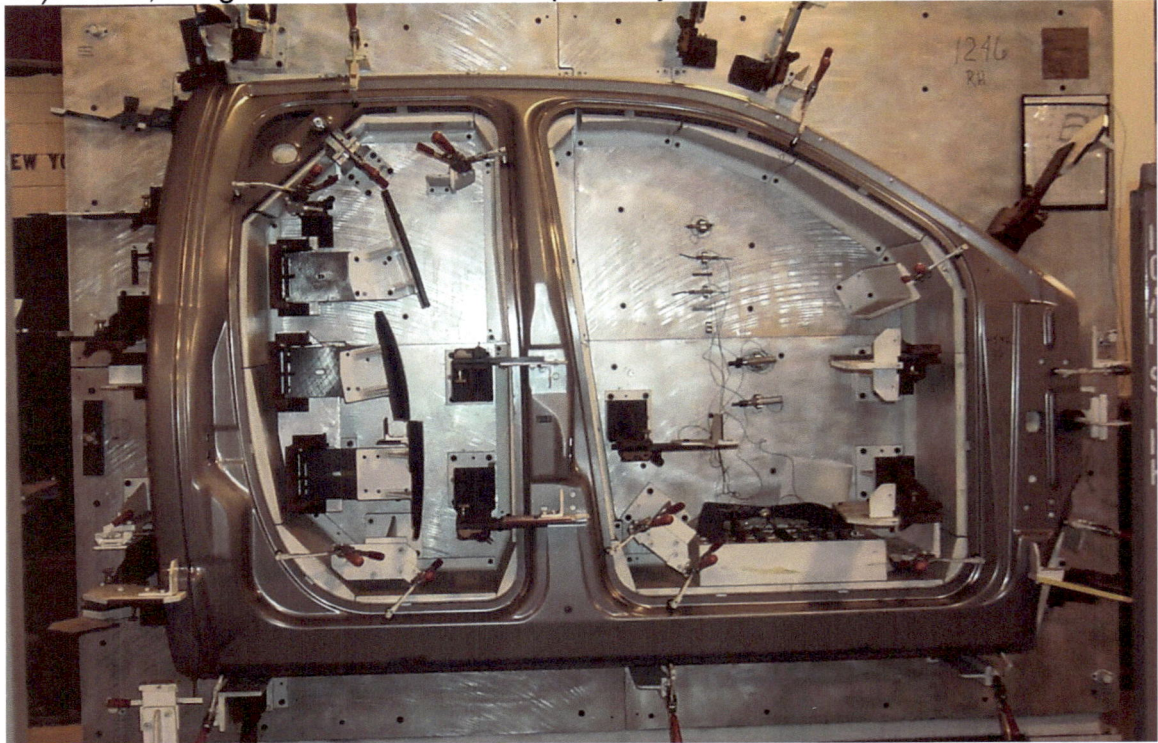

Figure 14 – Aperture outer detail in ring gauge

In this measurement gauge, represented in the above picture, the floorline in from body position (i.e. skin side out or facing the inspector). The panel is supported on 16 coordinated cross/car locators or PLP's (Principle Locating Points). The PLP's are coordinated in this fixture to match the assembly processing.

Figure 15 – Aperture Assembly marriage station

In this processing line, as represented above, the floorline is 90° from body position and skin side up (outer panel on top of inner panel). The panels are supported on 16 coordinated cross/car locators or PLP's (Principle Locating Points). Coordinated as the aperture inner and aperture outer panels have PLP's in the same location. In this fixture there is two sets of locator pins, one set for the inner panel and one set for the outer panel. The purpose of not utilizing the same holes between the two panels is that this method allows one panel to be adjusted to the other. If these panels were nested on the same pins, it would be impossible to make a shift of one panel to the other in the assembly tooling.

Another example of needing additional locating features, based on processing, is in a hinge to door assembly fixture. In this instance the door assembly is a rigid body requiring only (3) cross/car locators to create a plane and (2) pins (a 4 direction locator and a 2 direction locator). The three surface locators (cross/car) are positioned to create the largest triangle. One locator positioned at the front of the door near the top hinge, second one located at the upper b-pillar, and third one stationed at the lower b-pillar. This positioning is best suited for geometry of the part. However, with torque effects on the hinges, when bolts are tightened, an additional locator is required near the lower hinge. This locator is a functional "floating" locator, which means that it does not set the location of the door in the fixture. Once the door is located, this locator moves to the door and locks in place. This is necessary to maintain consistency in the door locators while not 'stressing' the door to a fixed location. In figure 16 below, you can see where the floating locator has been implemented.

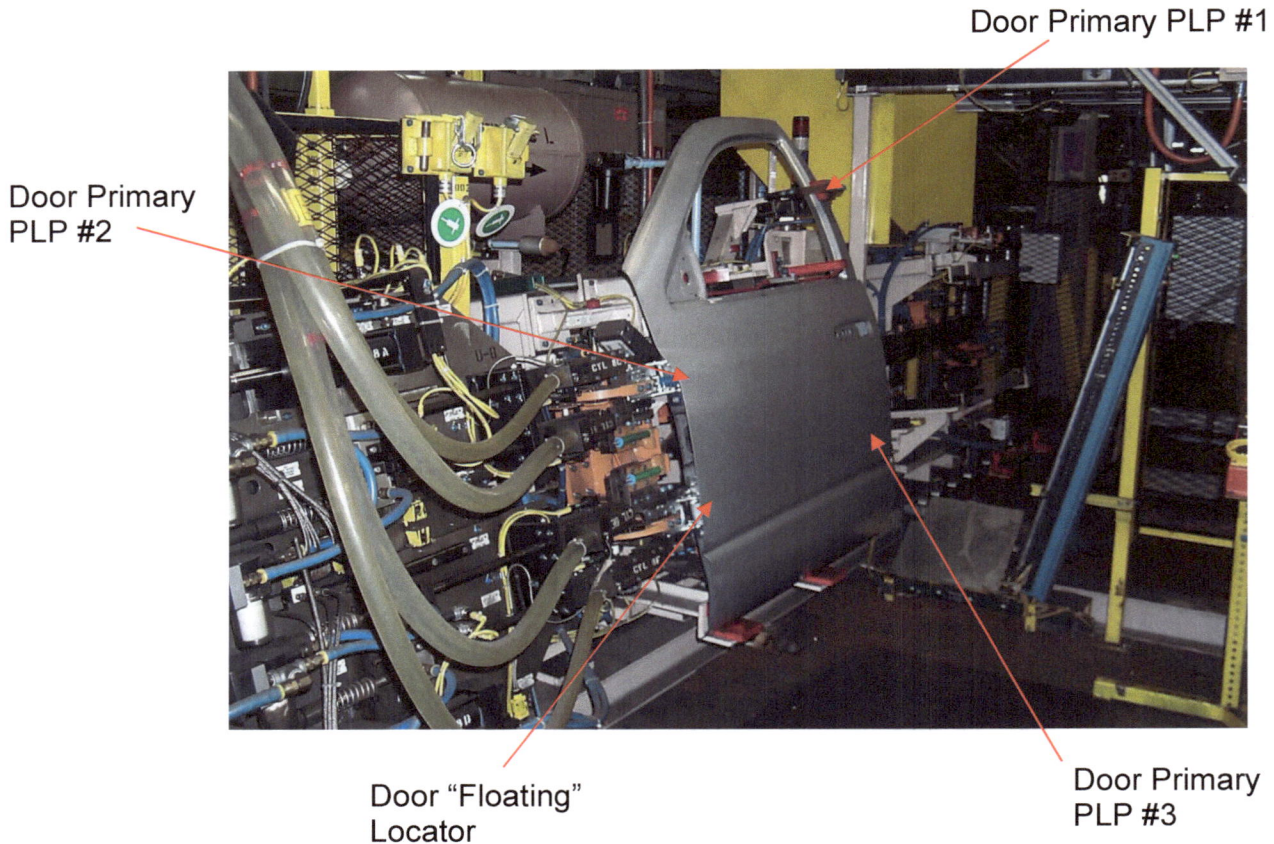

Door Primary PLP #1

Door Primary PLP #2

Door "Floating" Locator

Door Primary PLP #3

Figure 16 – Hinge to Door Assembly Fixture

In this fixture, note that primary PLP #2 is near the upper hinge that holds the door from moving or developing stress from the torque operation. Because no primary PLP is near the bottom hinge, a "floating" locator was added to reduce stress and/or torque effects on the door while the bolts for the lower hinge are tightened. Also note that while it is hard to see the locators or PLP's for primary PLP #2 & #3 and the "floating locator, it is due to the clamps for these PLP's swinging in from the back of the fixture through various access holes in the door inner panel. This allows the clamp finger to be set only a material thickness from the locator to hold the door in place. This method was not possible at primary PLP #1 so the clamp to locator pad distance is not easily maintained on the shop floor.

When datum selection and floorline cannot be adjusted to provide maximum support of the part, finite element modeling can also be used to determine the part deflection or anticipated deviation from nominal due to gravity constraints. Figure 17 through 22 are examples of one assembly in body position and "laydown" or 90 degrees from body with the differences in part deflection.

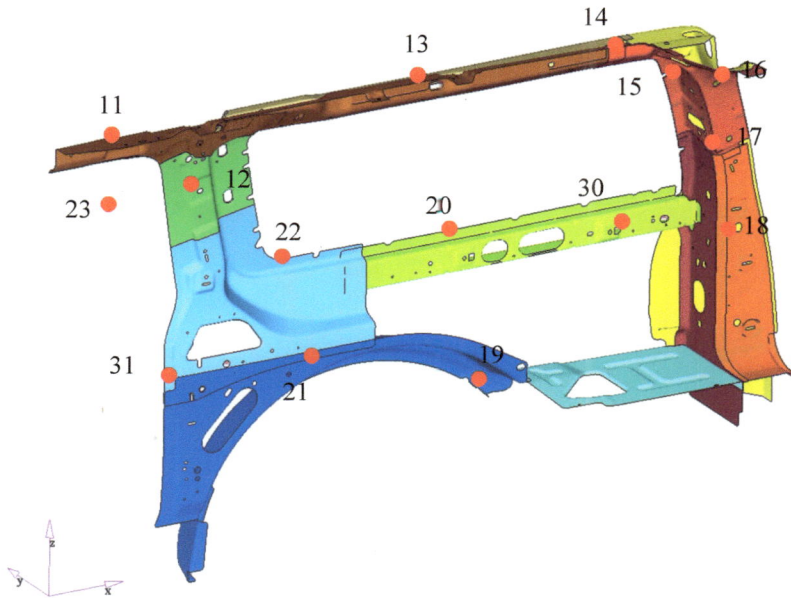

Figure 17
FEA model with key measurement points defined

The measurement points defined are coordinated with variable measurement points, in the various methods of verifying the part, either on a check fixture or gauge or through CMM measurement.

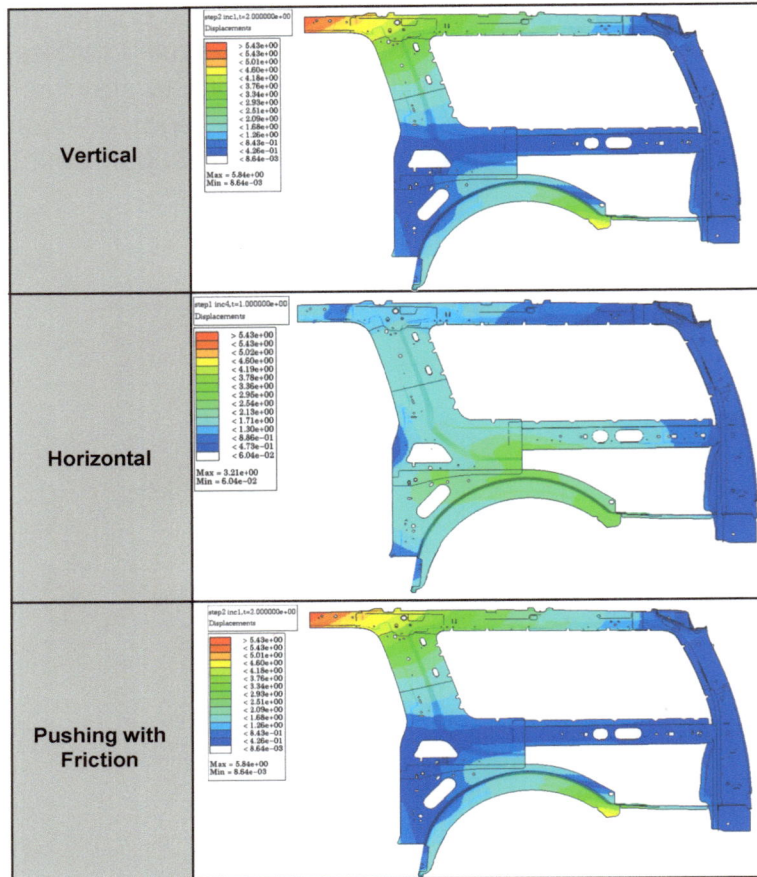

Figure 18
FEA Deflections – Three different studies of same assembly
1 – Vertical study: part in car position
2 - Horizontal study: part 90° from car position
3 - Vertical study: part in car position, friction study only

In figure 18, the various color maps indicate the amount of deflection expected due to gravity influences. The last color map is an FEA model on the amount of deflection caused by the friction of the part riding on the fixture or tooling pins. This model is in car position; thus, the gravity load and the coefficient of friction of the pins are applied.

1.5 g	FEA Point #	11	12	13	14	15	16	17	18	19	20	21	22	23	30	31
Disp. (mm)	Measured Points	1	2	3	4	5	6	7	8	9	10	11	12	13		
Vertical	AMP	4.771	2.583	2.294	0.847	0.753	0.635	0.271	0.239	1.259	0.491	1.276	0.633	3.211		
	Dx	0.437	0.014	0.411	0.589	0.598	0.619	0.260	0.219	-0.067	-0.040	-0.014	-0.047	0.150		
	Dy	-4.416	-2.552	-2.141	-0.596	-0.446	-0.134	0.079	0.100	-1.214	-0.400	1.259	-0.500	-3.110		
	Dz	-1.704	-0.343	-0.693	-0.118	-0.102	-0.045	-0.007	-0.001	-0.361	0.279	-0.203	0.381	-0.739		
Horizontal	AMP	0.965	1.379	0.908	0.396	0.365	0.360	0.206	0.214	2.273	1.425	2.619	1.767	1.489		
	Dx	-0.175	-0.143	-0.297	-0.368	-0.348	-0.358	-0.108	-0.117	0.160	-0.076	0.182	-0.094	-0.132		
	Dy	0.876	1.363	0.840	0.143	0.105	-0.016	0.005	0.022	2.172	1.382	2.537	1.731	1.481		
	Dz	-0.363	0.156	0.173	0.024	0.043	0.030	0.176	0.178	0.652	0.337	0.621	0.348	0.078		
Pushing w/Fric	AMP	0.034	0.021	0.041	0.025	0.022	0.028	0.028	0.078	0.370	0.715	0.444	0.428	0.034	0.927	2.183
	Dx	-0.033	-0.016	-0.040	-0.019	-0.014	-0.013	-0.008	-0.015	-0.051	0.031	-0.007	0.040	-0.030	0.023	-0.003
	Dy	0.004	0.008	0.007	0.016	0.014	0.015	-0.025	-0.075	-0.336	-0.714	-0.441	-0.420	0.010	-0.926	-2.183
	Dz	-0.007	0.012	-0.006	0.002	0.007	0.021	-0.009	-0.018	-0.147	0.000	-0.051	-0.068	0.013	-0.046	0.019

Figure 19 – Specific point results per FEA model

To achieve the proper results of the FEA model, it is important to ensure that the constraints are located in the proper spots with specific directional holds identified at the constraints. Figures 20, 21 & 22 show where these constraints were identified for each modeling position.

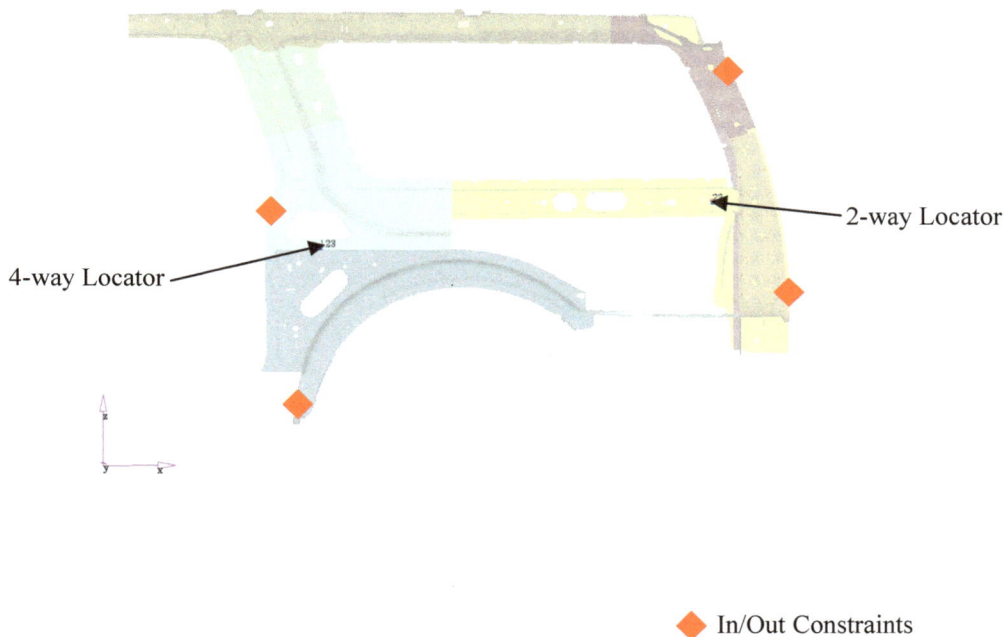

4-way Locator

2-way Locator

◆ In/Out Constraints

Figure 20 – Vertical FEA Model constraint locations

This figure depicts the constraints are chosen at net locations in rigid areas of the part. We did not use the nets at the top of the part to show how much the part will deflect prior to the clamps pulling the part to the net. Even though the clamp will pull the part, it does not necessarily mean that the part will move back into nominal position.

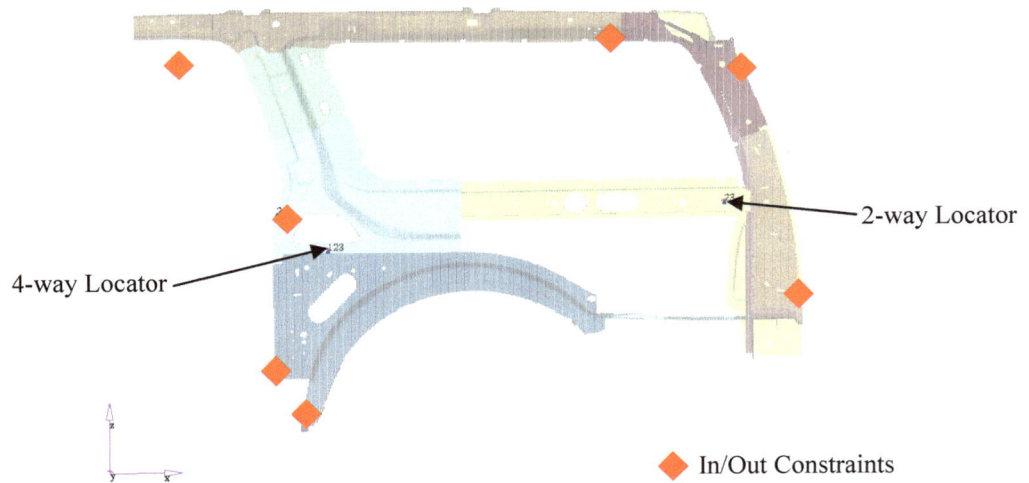

4-way Locator

2-way Locator

◆ In/Out Constraints

Figure 21 – Horizontal FEA Model constraint locations

Pictured above, the constraints are at all the net locations. In lay down position, the part will rest against all nets or the next corresponding part without need of clamping due to gravity effects.

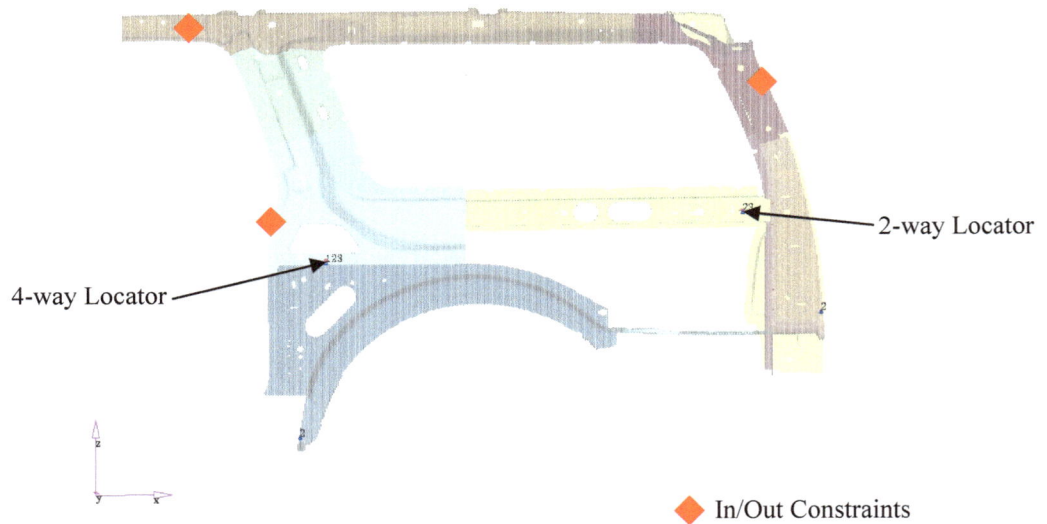

4-way Locator

2-way Locator

◆ In/Out Constraints

Figure 22 – Vertical Friction FEA model constraints

Here, the constraints represent the locations of where the operators will, by using their hands, load the part onto the fixture or tool. The purpose of this example is to show the amount of deflection near the locator pins due to part rigidity issues associated with gravity and friction. Note: While the part will not take a "set" in the deflected location after the part is removed from the tool or fixture, the clamps will not overcome the friction of the part on the pins when they are closed.

Another issue that needs to be addressed is related to part and any assemblies. One factor to consider is the measurement tolerances, or process specification limits, that actual part measurements will be compared against to determine if the part is acceptable or not.

The following methods of tolerance stack analysis focus on linear stacks, referring to one-dimensional (1D) models. While different engineering disciplines prefer certain types of tolerance stack analysis, an independent review of the various methods available is prudent to determine which stack method more closely represents a specific purpose.

Tolerance models correspond to either of two categories:
1. Roll up and roll down.
2. Loop.

Roll up and roll down rely on simple arithmetic plus take the form of tree diagrams. They can be used early in a product program as soon as preliminary designs and processes become available.

The loop category is more of a "hands on" study, as it provides a localized tolerance analysis than roll up or roll down studies. Both, of which, distribute tolerances equally throughout an assembly. Because the loop method focuses on tolerance allocation within a given part and process, it may be more reliable for the detailed work of tolerance adjustment within an assembly.

All linear tolerancing methods follow the same basic procedure. This sequence is as follows:
1. Define measurements (quantify objective)
2. Develop assembly sequence
3. Gather tolerance and process information
4. Perform calculations
5. Validate results
6. Review results with entire team

Regardless of which category is chosen, there are four different methods may be used to provide the final tolerance analysis:
1. Worst Case
2. Root Sum Squared (RSS)
3. Modified Root Sum Squared (MRSS)
4. Root Mean Squared (RMS)

The Worst Case method is the simplest and most conservative of all stack analysis. Derived by adding all the tolerances listed in the model, you will end up with the final result. Furthermore, it almost assures that all the components will fit together.

Classic Worst Case formula:

$$\text{Objective} = A + B + C + \ldots$$

The Root Sum Squared (RSS) method is an alternative method of achieving objective requirements when component and/or processing tolerances cannot be manufactured according to their design or Worst Case analysis.

Classic RSS formula:

$$\text{Objective} = \sqrt{t_1^2 + t_2^2 + \ldots t_n^2}$$

The Modified Root Sum Squared method is the same as the RSS method except that a correction factor is applied to the calculation. This is done to provide more accurate results and rely less on assumptions from the RSS method.

Classic MRSS formula:

$$\text{Objective} = \sqrt{(Cf)\, t_1^2 + t_2^2 + \ldots t_n^2}$$

The Root Mean Squared method (RMS) is a measure of magnitude of a set of absolute numbers (removing the negative signs from numbers in the set).

Classic RMS formula:

$$\text{Objective} = \sqrt{\frac{t_1^2 + t_2^2 + \ldots t_n^2}{n}}$$

Below is a list of pros and cons for each linear stack method. It can be used as an initial determinant of which stack is best for the application at hand.

Pros and Cons - Tolerance Stack Analysis Methods

Method	Pros		Cons	
Worst Case	1.	No assumptions on how individual component dimensions are distributed within the tolerance range.	1.	Large number of components mean large result.
	2.	Almost always guarantees that the components will fit together	2.	Small desired result mean small, often unattainable component and process tolerances.
	3.	Most appropriate for assemblies that require precise toelrance conformance due to safety or other critcal considerations.		
Root Sum Squared (RSS)	1.	Brings probability to tolerancing resulting in a more generous tolerance allocation.	1.	Probability assumes normal distribution
			2.	Assumes most components are manufactured near the center of the tolerance range rather than at the extremes.
Modified Root Sum Squared (MRSS)	1.	Similar to RSS method		
	2.	Applies a correction factor for more accurate results.		
	3.	Brings probability to tolerancing resulting in a more generous tolerance allocation.		
Root Mean Squared (RMS)	1.	Similar to RSS method	1.	Probability assumes normal distribution
	2.	Uses "measure of magnitude" to determine mean of components		

Category 1 stacks (Roll up and Roll down) can be created once the end result objective with tolerances is agreed upon, that tolerance is broken down (roll down) to the parts and processes that will be used to make the final product (see example in figure x). Once these tolerances are established, the entire team reviews them for buy-in. This exercise enables all parties to understand what requirements are needed by their individual piece of the pie. Another way to use category 1 stacks is to roll up. This process that takes known processes or product tolerances, and based on the established tree, rolls up the tolerances to a final objective.

Category 2 tolerance stacks are based on data gathered from previous production processes and parts and either 3D modeled or Linear stacked (rolled up). This method then gives the final result that is compared to the objective to determine if, with these particular processes for creating the part and assembling it, can meet customer expectations. If the stacked or "rolled up" value is larger than the objective, then the team must decide where to reduce tolerances by improving a process to achieve the objective. Below, figure x, is one example of a Linear Stack Analysis. In this example, the RSS (Root Sum Squared) result is used.

Door to Door - c/c direction

Study Date:

Study#

Author:

Item:	System Description	Tolerance Description	Range	± Stack	RSS Calc
1	Aperture hinge mounting surf - frt. Door	Profile	1	±0.5000	±0.2500
2	Aperture hinge mounting surf - rr. Door	Profile	1	±0.5000	±0.2500
3	Rr. Door Hem edge in door assembly	Profile	1.5	±0.7500	±0.5625
4	Rr. Door assy. C/c PLP's in hinge fixt.	Tool Block	0.625	±0.3125	±0.0977
5	Rr. Hinge c/c PLP in hinge fixture	Tool Block	0.625	±0.3125	±0.0977
6	Frt. Door assembly c/c PLP's in door hang tool	Tool Block	0.625	±0.3125	±0.0977
7	Frt. Door Hem edge in door assembly	Profile	1.5	±0.7500	±0.5625
8	Frt.Door assy. C/c PLP's in hinge fixt.	Tool Block	0.625	±0.3125	±0.0977
9	Frt. Hinge c/c PLP in hinge fixture	Tool Block	0.625	±0.3125	±0.0977
	Results		8.125	±4.0625	±1.4537

Figure 23 – Linear Stack Analysis example

In the above figure (#23), the linear stack analysis example is the flush objective for front door to rear door. A linear stack analysis does not include nominal locations for each part, only the assigned tolerances for each part and process. This stack, RSS results of +/-1.45mm, indicates that the parts and process will meet the objective requirement of +/-1.50mm.

Using the same tolerances from above, we can also use the SAE (Society of Automotive Engineers) benderized method. This method attempts to account for "mean shifts" in the variables as the mean of a variable does not always have a 0 nominal deviation.

Door to Door - c/c direction
SAE Benderized method

Study#

Author:

Item:	System Description	Tolerance Description	Range	± Stack	RSS Calc	Mean
1	Aperture hing mounting surf -frt. Door	Profile	0.8	±0.4000	±0.1600	0.200
2	Aperture hing mounting surf -rr. Door	Profile	0.8	±0.4000	±0.1600	0.200
3	Rr. Door Hem edge in door assembly	Profile	1.3	±0.6500	±0.4225	0.200
4	Rr. Door assy. c/c PLP's in hinge fixture	Tool Block	0.625	±0.3125	±0.0977	
5	Rr. Hinge c/c PLP in hinge fixture	Tool Block	0.625	±0.3125	±0.0977	
6	Frt. Door assembly c/c PLP's in door hang tool	Tool Block	0.625	±0.3125	±0.0977	
7	Frt. Door Hem edge in door assembly	Profile	1.3	±0.6500	±0.4225	0.200
8	Frt. Door assy. c/c PLP's in hinge fixture	Tool Block	0.625	±0.3125	±0.0977	
9	Frt. Hinge c/c PLP in hinge fixture	Tool Block	0.625	±0.3125	±0.0977	
	Results		7.325	±3.6625	±1.2858	0.800

.25ET	1.83
RSS	
.75 RSS	1.93
Variation = .25 ET +.75RSS	3.76
Variation + Shift	**4.56** **±2.28**

Figure 24

Here, in figure 24, you can see that by trying to account for any mean shift in the variables, the anticipated overall tolerance can increase from +/-1.45mm (standard RSS stack) to +/-2.28mm (benderized RSS stack).

While there is no steadfast rule on when to use the standard RSS stack or the benderized RSS stack, it is recommended that where you have multiple parts going into a stack, the benderized stack method provides better correlation to the actual results you may see.

Loop Diagram

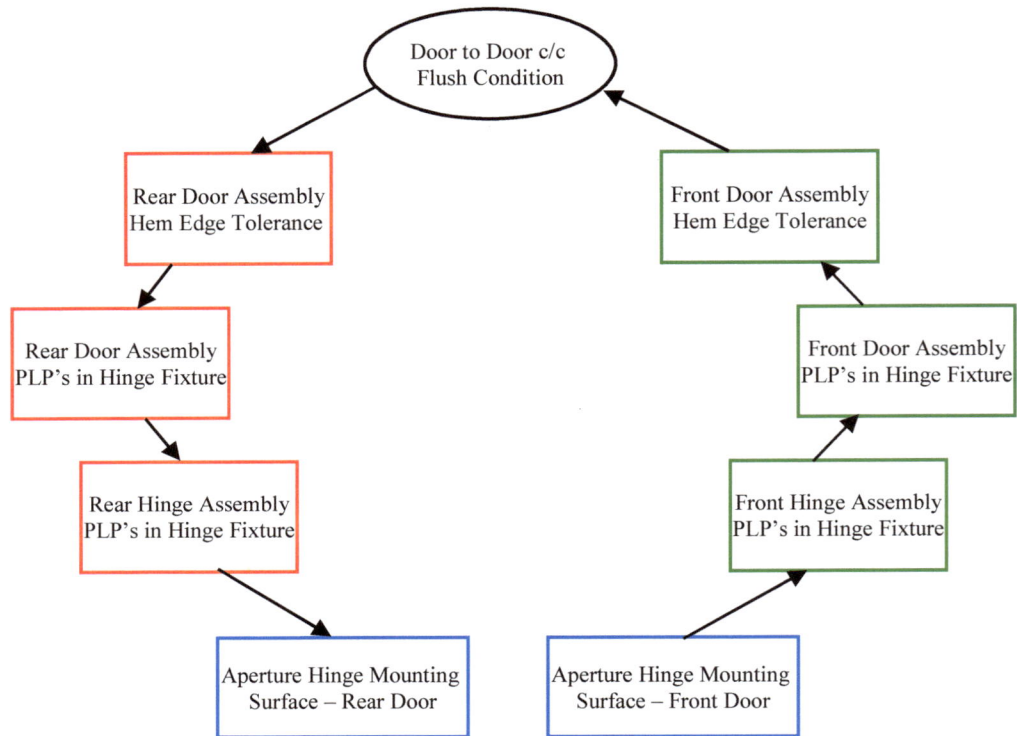

Figure 25

Figure 25, above, reflects a loop diagram. The loop diagram starts with the objective, in this case, the door-to-door cross/car flush condition. It starts at one part (rear door assembly) looping around, to the mating objective part (front door assembly), by stepping through each part and process tolerance that is needed to complete the assembly of door to door on the vehicle. Here, starting from the objective, the rear door assembly has a hem edge tolerance to its' PLP's. Next, is the processing tolerance of the rear door assembly PLP pads in the rear door to hinge assembly fixture. Afterwards is the processing tolerance of the hinge PLP pad in the door to hinge assembly fixture. The part tolerance for the aperture outer rear door hinge-mounting surface is next. Final phase involves the parts and processing of the front door (are the same as the rear door); the other half of the loop is completed in reverse order for the front door.

Tree Diagram

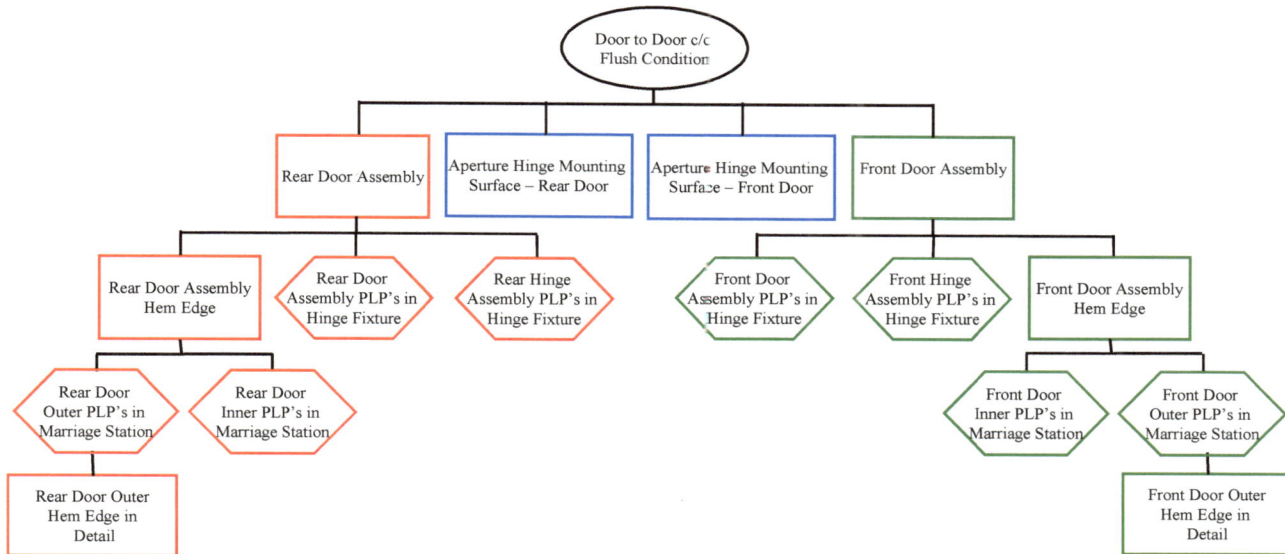

Figure 26

Above is a tree diagram for door-to-door flush. As you can see, the tree diagram gets its name due to all the "branches" or levels. Each level represents one phase of the build or tear down, of the complete objective, to the details themselves. This diagram, mostly used in roll up or roll down stacks, is usually generated initially with only the first one or two levels created. This is done to reduce stack setup time and to get an idea of what kind of tolerance is available for each part and process. At this time, in a production cycle processes, for the lower levels may not yet be defined making it difficult to create an accurate tree or produce accurate results at these sub-levels. Additionally, there is no benefit to continuing the stack to the lower branches if the initial branches have unachievable tolerances.

Advantages and disadvantages of 3D model variation analysis, compared to a Linear Stack analysis, is that the linear stack analysis is two-dimensional only and does not require a completed math model. This allows the linear stack to be completed earlier in the product timeline. The 3D model also requires a generated "tree" or step procedure on how the part is processed up to the point of the objective. Advantages of using a 3D model offers the opportunity to show part mating issues that might not be found in a linear analysis. Why? Because it is an all directions analysis. In the 3D model, the requested end product capability desired can be input (Cp =1.33 or 1.00 or 1.67,etc.) and end results shown. Also with the 3D model, an animated mock-up is created so the team can view how the part is processed, step-by-step, until the final result is achieved. An example of this type of variation analysis is shown in figure 27.

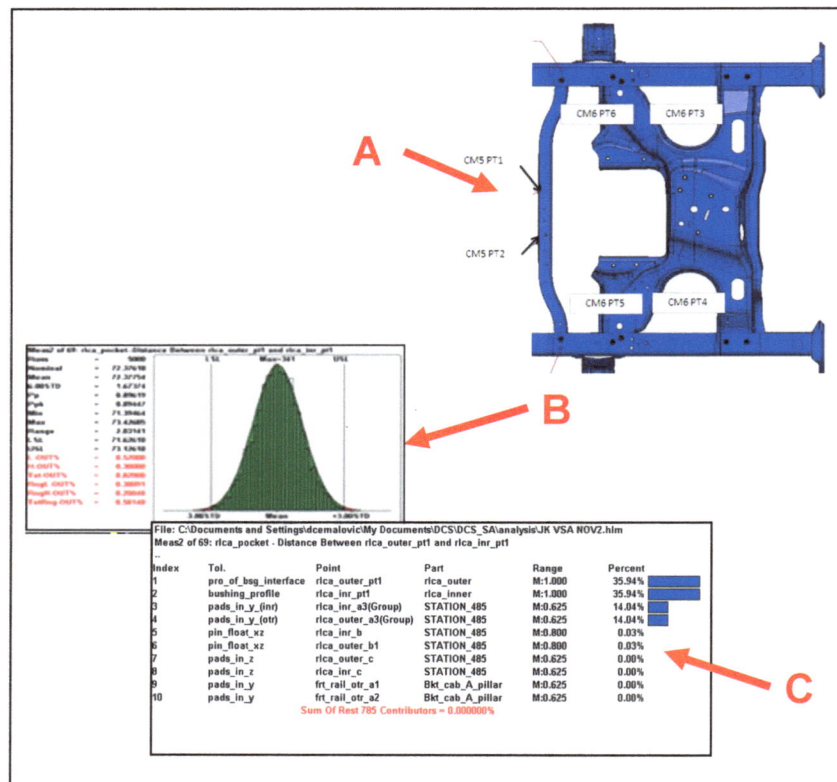

Figure 27

Figure 27 is a typical 3D variation simulation, in solids mode, report showing the points in the model (A), the histogram of all tolerances (B), and the percentage each tolerance contributes to the overall reported variation (C).

Pre-Production (Pilot) Phase

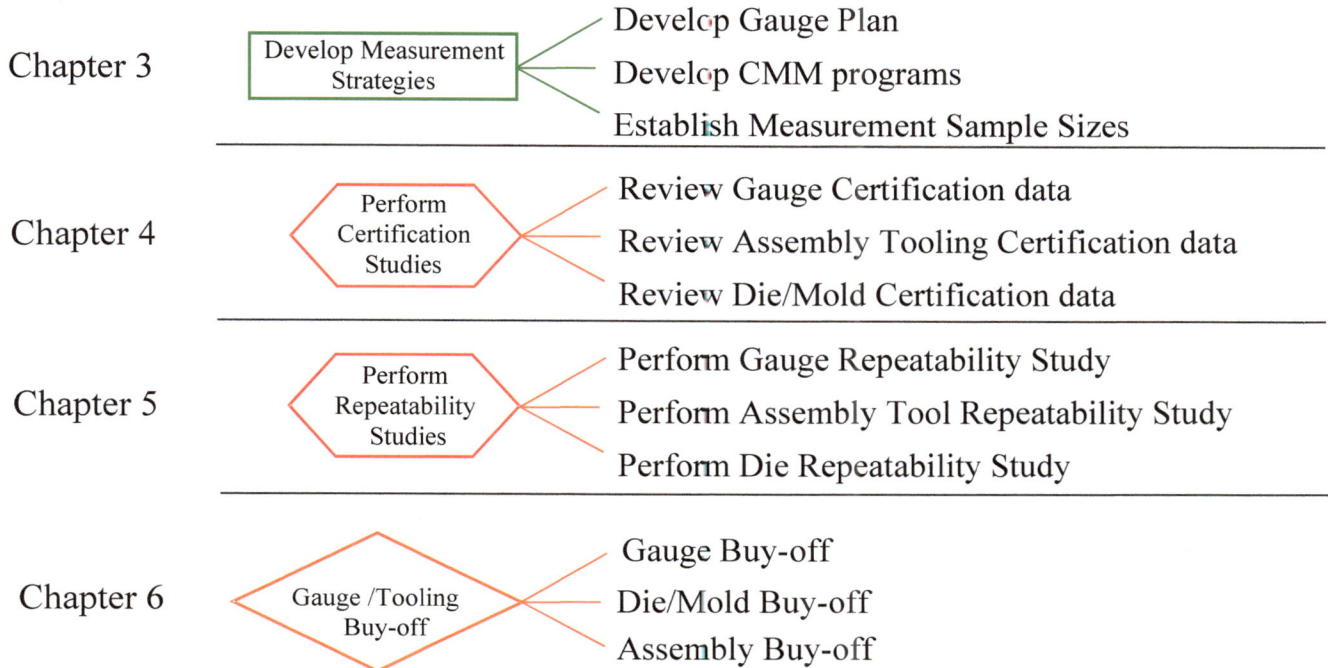

Chapter 3   **Develop Measurement Strategies**
- Develop Gauge Plan
- Develop CMM programs
- Establish Measurement Sample Sizes

Chapter 4   **Perform Certification Studies**
- Review Gauge Certification data
- Review Assembly Tooling Certification data
- Review Die/Mold Certification data

Chapter 5   **Perform Repeatability Studies**
- Perform Gauge Repeatability Study
- Perform Assembly Tool Repeatability Study
- Perform Die Repeatability Study

Chapter 6   **Gauge /Tooling Buy-off**
- Gauge Buy-off
- Die/Mold Buy-off
- Assembly Buy-off

Chapter 3 - Developing Measurement Strategies

In developing a measurement plan, one needs to understand the resources available and identifies "what is hoped to be gained", such as: costs for CMM (Coordinate Measurement Machine) programming, gauge creation, gauge and program prove-out, along with actual measurement sample sizes. All these factors will have a great impact on a product budget.

Before getting into the specifics, it is important to understand and quantify two vital questions, "What is to be gained by measurement" and "Where do you want to gain this information". In most cases, you will want to measure as close to geometry set operation(s) as possible. The objective is to give you an understanding of the capability of the geometry station without the influence of additional processes. It also helps conclude whether or not "enough welds are in the station to set geometry" or whether three screws hold geometry versus four. Having stated this, a clear understanding must be that while in pre-production (or prior to launch) timing may afford the ability to measure after each geometry station. Measuring at these places during production may be limited to part availability, due to geometry station in middle of process line, or CMM/gauge accessibility.

Measurement plans start by determining which phases of the build measured will produce data to show capability of the various processes and provide value for root cause analysis. Obviously, it is essential to measure the complete product in order to provide overall confidence in the products' manufacturability for you and your customer.

The next place to go is to any major geometry set stations, then finally down to the individual details. Once this segment is established, the subsequent phase involves how they will be measured: a full ring gauge, CMM with a holding fixture, CMM utilizing flexible fixturing, or another method.

Developing CMM programs

In developing a correct CMM program the following items are critical in determining if, after a measurement is complete, the data is valid:

1. Point Names - these are the point ID's that best describe on what part the point is measured, the type of point, and the direction of measurement.
2. Point Location – the actual nominal of the measured point, commonly known as the X, Y, and Z-axis location.
3. Point Vector – the angle of the surface, which is the normal direction that the CMM probe approaches the point to acquire the most accurate measurement.
4. Part Compensation – in some cases, the point identified as "design side" in the CAD model, may be on the opposite side the measurement is actually taken on. In these instances, the part thickness must be added to the program so that the program will offset the measurement with that value.
   Note: When adding part compensation, to the CMM program, it should be added in an area where the nominal point location will not be altered by the calculated part compensation value. In PCDMIS, there are two ways to enter part compensation. As an actual compensation or as a theoretical compensation. The actual compensation will keep the point location values as they are with respect to the CAD model. The theoretical compensation will adjust the point location to the measured side of material.
5. Tolerances – the actual tolerance of the part at the point location. This tolerance is used in calculating all statistical information of the measurement sample and in generating the limits for run charts.
6. Measurement Alignment – this is how the CMM will align to the part coordinate system before actual measurement. Some methods are:
   a. Interative alignment – this is where the CMM aligns to the setup locations on the fixture, then measures the locating scheme nets and pins for fit. This alignment is only repeated prior to each sample grouping measurement.
   b. Part alignment – this is where the CMM aligns to the part by "finding" the locating features on the part, the 4way hole, 2way slot, net locations, etc. and after measuring them, generates a "best fit" alignment to that part. This step is repeated for each part measured.
   c. Fixture alignment – where the CMM aligns to the CMM holding fixture setup features (tooling balls, setup blocks, etc.), and best fits to the coordinate system.

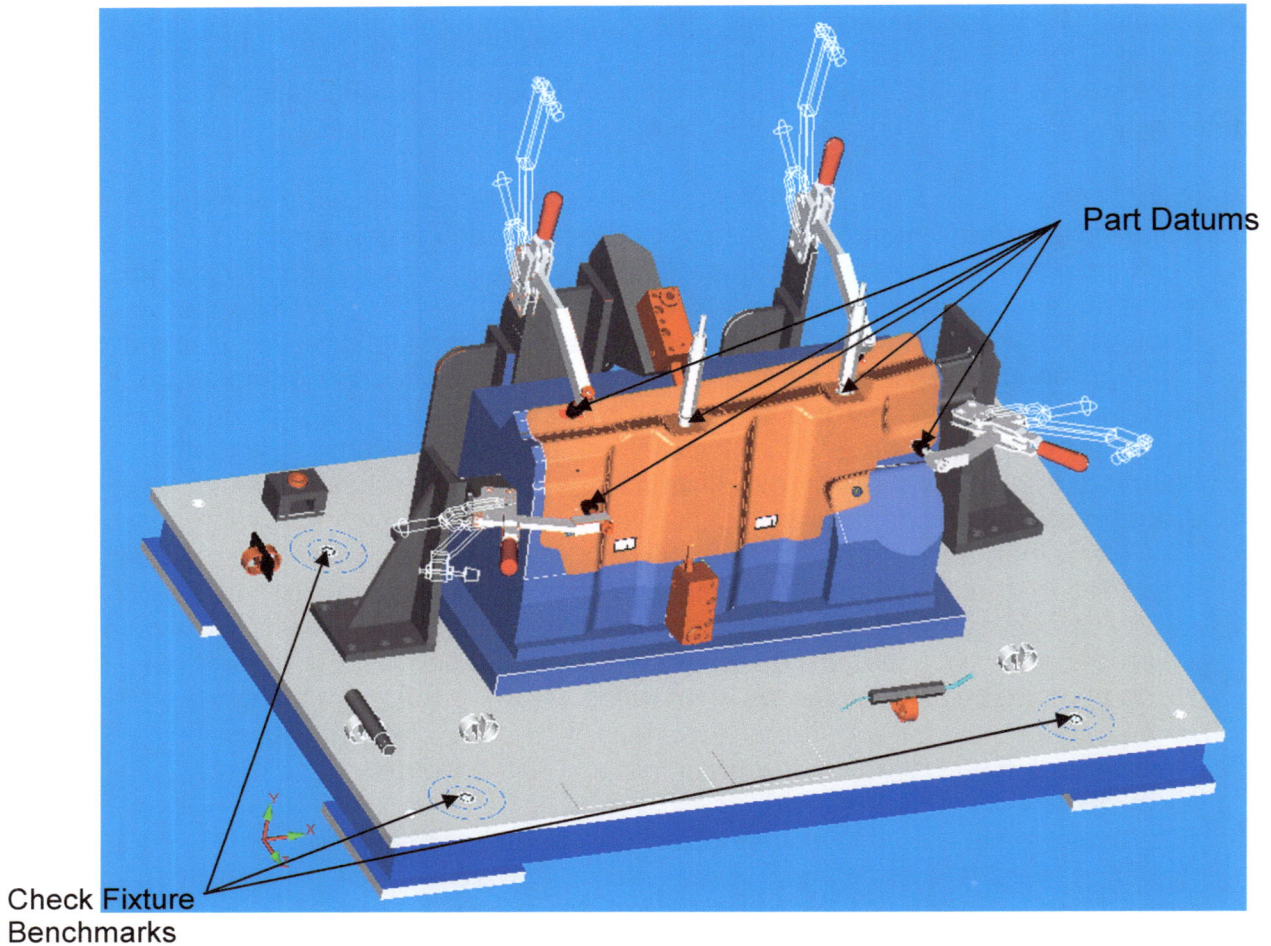

Figure 28

In figure 28, there are two ways of aligning to measure this part. One way is to use the fixture benchmarks, which sets a 3-2-1 alignment. The second way is to use the part datums. The above figure represents a 4-2-1 alignment and will require the locator pins to be removed prior to part alignment; but, after the part has been clamped, so that the CMM can measure the locator holes.

While aligning, to the part, removes the fixture tolerance associated with the datum build certification back to the benchmarks. Removing of the locator pins may cause the part to move, additionally, as with this example, the 4 "surface datums" are not planar. Thus, creating error in the alignment method. This is due to the fact that the two upper datums, near the locator pins, can only have one direction associated to the plane it is creating. Since they are on an angle, the outcome may dictate, either a loose control in the other direction or "fight" the direction established by the hole locators.

Developing Measurement Strategies – Pre-Production

In an effort to learn as much as possible about detail components and the assemblies they make prior to product launch, some key factors have to be considered:

1. How many detail parts will be made? How many batch runs will the detail parts be made in?
2. How many assemblies will be made? Is there opportunity to change the assembly tooling during build?
3. How much time is available to measure?
4. What resources are available for measurement?

But, most importantly identify, "what do we expect to learn or gain by this effort"?

At this point in time, it is also ideal to learn about the measurement devices themselves, which you will be using during production. For this reason, gage repeatability studies must take place on all measurement devices before any part/assembly measurement begins. In most cases, sufficient part quantities will not be available to perform a GR&R at this stage; so, a GR will have to suffice. Below, we will attempt to answer the questions mentioned above. In some cases, we will provide examples to support our answers plus present an idea of when (in a product timeline) this step may take place.

1. ***How many detail parts will be made?*** How many batch runs will the detail parts be made in? The measurement sample size will be based in part by the number of detail parts and final assemblies ordered and in what quantity batches they are sequenced. It is important to understand the detail part variation, over the entire pre-production cycle, even if measurement is limited to a one time only opportunity. In most cases, detail parts are often generated from a single batch run. This is primarily due to die/mold set up costs needed for multiple batches. One other factor that may have significant effects, on detail part variations, is the amount and type of shipping containers the parts will be transported to the pre-production build source in. Amount of containers will determine batch run size or where the parts may be stored temporarily. All these factors will have an influence on the measurement data. Dimensional measurement, at this stage, is just as important as: weld certification, process prove-out, and building units for safety testing, etc. With this in mind, measurement plans need to be developed as to not curtail other events from happening. Having said all of this, however, a base line for measurement should be 30 samples per batch. While 30 samples per batch is the goal, 30 measurements may need to be spread evenly to the amount of part batches produced to give an understanding of overall part quality. Additionally, costs associated with measuring more than 30 samples may prove to be an ineffective use of monies. While these parts are mostly from non-production methods (prototype dies, hand formed, etc.), it is important to understand the variation within the parts. This process will assist in root causing

assembly build issues and provide an understanding of what assembly variation/mean deviations is attributable to the details.

2. ***How many assemblies will be made?*** Is there opportunity to change the assembly tooling during build? In the ever-tightening timing of a product to market cycle, time to generate pre-production units for measurement and testing is extremely tight. One must consider what assemblies are important to measure, since time usually does not permit all being measured, and what characteristics on these assemblies provide the most knowledge of the process when measured. This information is also important since this same list of measurement assemblies should be used during production life of the product. Again, at this stage, a base line of 30 samples should be measured. This sample size provides a good confidence level in the actual numbers reported. It is important to note that 30 samples can be measured in subgroups totaling 30. For example, 6 groups of 5 samples as long as there is confidence that no change to any part of the process has occurred between each subgroup measured. This method is generally accepted due to measurement time constraints, storage issues, and pre-production build schedules commonly experienced. If changes are possible in the assembly tooling during pre-production build, measurement batches should be limited to sample sizes of 5, until the measurement data indicates a process requiring no further change. At this point, 30 samples should be measured for process capability. Here again, these 30 samples can be broken into sub-groups of 5 for measurement purposes. Also, compiled for process capability values as long as no changes have been made to the process between sub-group measurements; otherwise, the process of measurement must be restarted.

3. ***How many detail parts will be made?*** How many batch runs will the detail parts be made in? The measurement sample size will be based in part by the number of detail parts and final assemblies ordered and in what quantity batches they are sequenced. It is important to understand the detail part variation, over the entire pre-production cycle, even if measurement is limited to a one time only opportunity. In most cases, detail parts are often generated from a single batch run. This is primarily due to die/mold set up costs needed for multiple batches. One other factor that may have significant effects, on detail part variations, is the amount and type of shipping containers the parts will be transported to the pre-production build source in. Amount of containers will determine batch run size or where the parts may be stored temporarily. All these factors will have an influence on the measurement data. Dimensional measurement, at this stage, is just as important as: weld certification, process prove-out, and building units for safety testing, etc. With this in mind, measurement plans need to be developed as to not curtail other events from happening. Having said all of this, however, a base line for measurement should be 30 samples per batch. While 30 samples per batch is the goal, 30 measurements may need to be spread evenly to the amount of part batches produced to give an understanding of overall part quality. Additionally, costs associated with measuring more than 30 samples may prove

to be an ineffective use of monies. While these parts are mostly from non-production methods (prototype dies, hand formed, etc.), it is important to understand the variation within the parts. These processes will assist in root causing and making sure that the person measuring is the same all through that measurement phase. Or, that the new person is just as repeatable as the previous person. Time scheduling should include GR's, sub-groups of measurement needed, various assemblies to be measured, and what system setup requirements are needed for each assembly to be measured. Examples include: fixture change-outs, probe tip changes, calibrations, etc.

4.   ***What resources are available for measurement?*** Ideally pre-production resources, both system and personnel, should be the same for manufacture (production). In cases where personnel are needed for measurement execution, such as actually taking the measurements themselves, having production personnel available at pre-production allows for training to be completed, and any measurement issues to be resolved prior to production. As far as system resources, utilizing the same equipment from pre-production to manufacture will help in developing part/assembly-holding techniques, measurement schedules, and reduce any unforeseen measurement issues once the product is placed at the production run source.

Shown in Figure x diagram below, indicates what parts are needed for the entire product along with a flow diagram for the assembly process. You can also see colored boxes that will be shaded based on status of gage repeatability, tool repeatability, part capability, and assembly capability requirements. Common shading practices throughout industry standards:
- **Green** – passed particular requirement.
- **Yellow** – failed requirement, plan in place to re-test.
- **Red** – failed requirement; currently no plan in place to correct or re-test.

Additionally, figure x also provides an example of the ordering quantities per week that includes pre-production build and the amount of assemblies to be built.

In this section, thus far, we have focused on pre-production dimensional measurements, and wherever possible its relationship to production dimensional capacity. We also identified other quality items that need to be measured, such as: welds (location and integrity), torque, water testing, etc. Based on these processes, we learned that these items can also be monitored statistically and should be performed in pre-production similar to production methods.

Furthermore, we explored other additional production dimensional measurements that needed to be considered at this stage are any in-line systems. In-line meaning that the measurement system is part of the production assembly line. These systems can be in-geometry or end-of-line depending on the situation, line space, etc. It is important to note that when determining which system location will be used, in-geometry or end-of-line, a list of resource requirements needs to be created and addressed adequately to ensure full potential of the measurement system. Some of these resource requirements include:

- In-geometry
    - Station cycle time constraints
    - Who is responsible for maintenance
    - Retractable pins and nets not on main assembly
    - Measurement system calibration

- End-of-line
    - Who is responsible for maintenance
    - Measurement system calibration
    - Additional floor space requirements (if separate station)

In either system, additional issues that need to be addressed include:
- Personnel training to review data
- General data storage server with access to all plant personnel
- Budget for upkeep of the system
- What capability will this system have to the overall build cycle? Stated another way, will this system be capable of stopping the production line if non-conformances occur?

Another consideration for this type of measurement is the actual measurement system itself. There are a wide variety of systems available in the marketplace, some commonly used dimensional resources include: vision, photogrametry, CMM, white light scanning, laser scanning, etc. Each has specific capabilities, system requirements and cycle time constraints. In most automotive manufacturing environments, the vision system is used for 100% in-line inspection. Vision systems, an example shown in figure 29, are individual laser sensors mounted on tubular frames that measure specific points on a product. Each sensor has a specific "field of view" that is limited by the triangulation that is generated from the laser to the view window.

Vision Sensor

Tri-Light

Figure 29 – Standard Vision Station

In this particular vision system, sensors w th tri-lights are mounted on a tubular frame. For part alignment, a lifter raises the vehicle off of a carrier holding the vehicle by the datums specified. In this case the vehicle is held in a 4-2-1 alignment meaning 4 up/down net pads, a 4-way pin for cross/car and the fore/aft directions, and a 2-way pin for a cross/car anti-rotate.

Figure 30

Another example of in-line measurement is in figure 30, above. Here the laser sensors are mounted on an end-effector, which is manipulated into position by a robot. These sensors measure the rear window opening and feed that information back to the robot's program which then adjusts the position of the end-effector until the rear window can be installed into the center of the opening.

Photogrametry systems are cameras that can be mounted on a robot. Photogrametry relies on the ability of the camera taking many 2D digital pictures with a reference (alignment) system. It has the capability to reassemble all the pictures, creating a 3D file where data is stored, based on the targets the system is measuring.

The next step involves defining measurement point locations. Once the points are established, they will be used for two purposes:
- Initial part or assembly buy-off
- Continuous part or assembly monitoring

The dimension points are usually designated two ways. "Non-critical" for points that will only be used for initial part or assembly buy-off. "Critical" for points that will be used to monitor part or assembly quality throughout the life of the product.

Measurement points should be limited, though not entirely, to around 30 points per panel critical and non-critical inclusive. This general rule of thumb is to help determine and schedule CMM programming time, measurement time, fixturing expenses, etc. While 30 points should be the limit some parts, an aperture outer panel and door assembly, for example, have measurement points at customer audit locations. Additional points for seal gap surfaces, position for hinge mounting surfaces, and all mating part surfaces and holes. Because of this situation, aperture panels and door assemblies usually average around 90 points.

When determining which points are critical versus non-critical, the first step is to identify any process or safety critical areas. These surfaces may have 3 points each to show angle of the plane they make, but only 1 or 2 will be critical for part/assembly process monitoring (see figure 31). Previous part datums should be critical, meaning that it is a locating feature in an assembly tool that holds a smaller part to the larger part. But, the locating feature that is not part of the larger parts main datuming scheme (see figure 31). Other measurement points should include at least 3 points for each geometry procedure even though some may be the same from process to process.

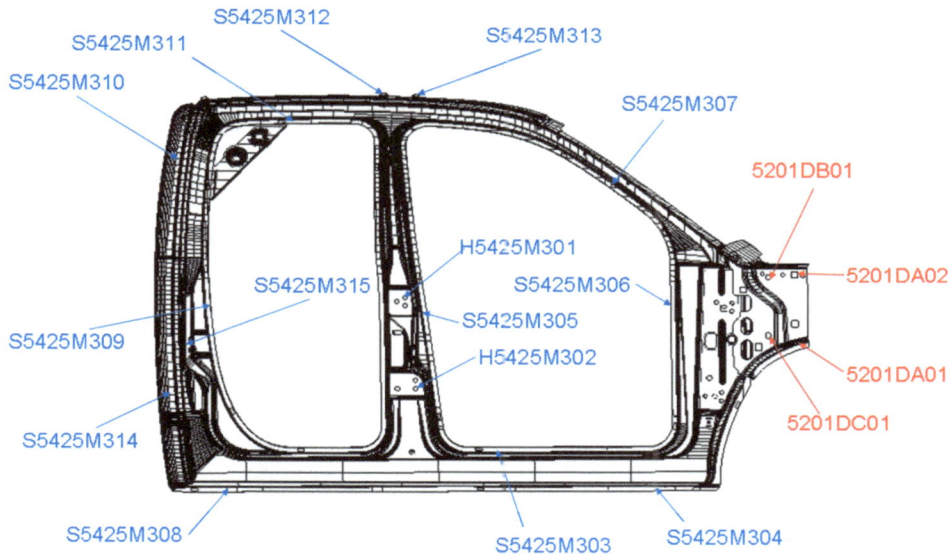

Figure 31 – Example of Measurement Points

In figure 31, the points identified as "D" (DA01, DA02, DC01, and DB01 in red) are the datum points for the smaller detail (part #5201) that goes to the larger detail (part #5425).

Once these points are all generated, they will need to be allocated as to what they will be used for. Some attributes to consider are:
- GR & GR&R studies
- Vision system measurement
- Assembly process measurement
- Part/Assembly audit locations
- Critical characteristic

In addition to the above measurement point attributes, some datums may also need to be attributed based on how the part/assembly will be held and aligned to for measurement. This situation is necessary in "flexible" fixturing systems (see figure 32). Here, only some of the datums are used to hold the part due to system constraints whereas the remaining portions are used to align to the part. In the case of an aperture assembly, 16 cross/car datums are needed for the assembly tools from detail to complete assembly (see figure 33). However, for measurement purposes, only 8 holding supports are needed and only 4 datums will be used to create the cross/car plane (see figure 33).

Figure 32 – Example of flexible fixturing

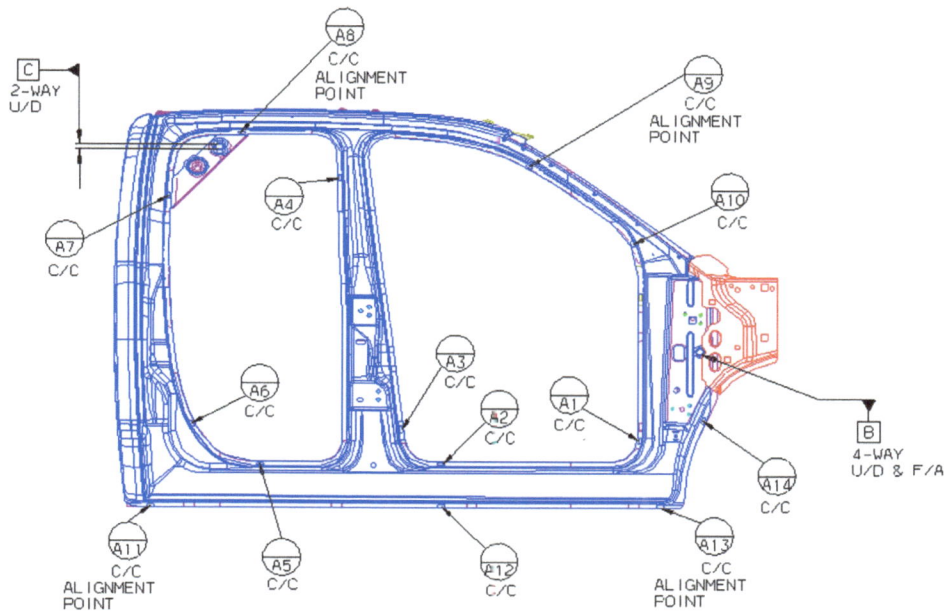

Figure 33 – Datums and Alignment points

Point naming conventions vary from company to company, though some key items should exist. Point names should reference the part number (while not the part number in entirety, but some portion so that the downstream users can distinguish what part a measured point on a measurement report is reflecting). The point name should also distinguish whether the point is a datum point, a measurement point, or an audit point. This information is helpful when determining how to make adjustments in the tooling to correct build issues. Another item that needs to be captured in the naming convention is the part attribute. For example, is the point on a surface? Located in a hole? Identified on an edge? Or, is it considered a gap point, flush point, or a sealgap point? With many of these questions answered in the point name, almost anyone can understand, to some extent, what a particular point is measuring (see figure 34).

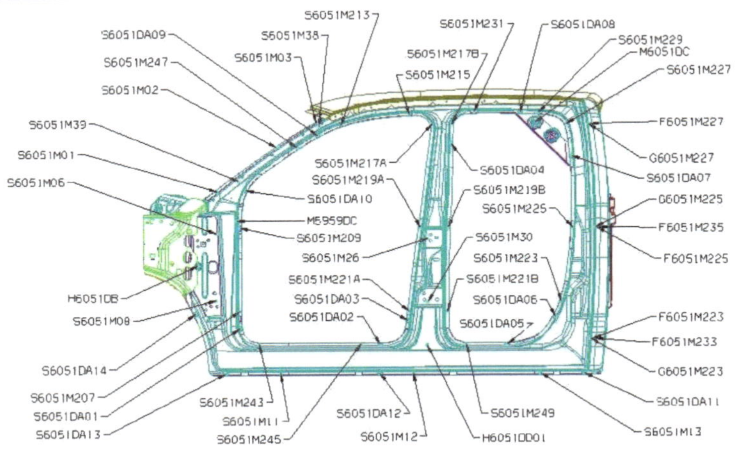

LH QUAD APERTURE ASSEMBLY COMPLETE MEASUREMENT POINTS

Labels (graphic):
S6051M213, S6051M231, S6051DA08, S6051DA09, S6051M38, S6051M229, M6051DC, S6051M247, S6051M03, S6051M217B, S6051M227, S6051M02, S6051M215, S6051M39, F6051M227, S6051M01, S6051M217A, S6051DA04, G6051M227, S6051M06, S6051M219A, S6051M219B, S6051DA07, S6051DA10, S6051M225, G6051M225, M5959DC, S6051M30, F6051M235, S6051M209, S6051M223, F6051M225, S6051M26, S6051M221A, S6051M221B, H6051DB, S6051DA03, S6051DA06, S6051M08, S6051DA02, S6051DA05, F6051M223, S6051DA14, F6051M233, S6051M207, G6051M223, S6051DA01, S6051M243, S6051DA12, S6051DA11, S6051DA13, S6051M11, S6051M249, S6051M245, S6051M12, H6051DD01, S6051M13

CMM PROGRAM: LH QUAD APERTURE ASSEMBLY

POINT NAME	X Y Z	I J K	AXIS TOLERANCE	CAD MODEL TOLERANCE	REPORTING (Y OR N)	HOLE NOMINAL	HOLE TOLERANCE	SLOT WIDTH NOMINAL	SLOT WIDTH TOLERANCE	SLOT LENGTH NOMINAL	SLOT LENGTH TOLERANCE	ALIGNMENT POINT	AUDIT POINT	VISION SYSTEM REPORTING (Y OR N)	VISION SYSTEM X Y Z	VISION SYSTEM I J K
S6051DA10	2102.198	0.015	0.02		N											
	-838.718	0.973	1.17	1.20	Y											
	1850.000	-0.229	0.27		N											
S6051DA11	3861.000	0.000	0.00		N											
	-821.278	0.966	0.25	0.25	Y							Y				
	864.425	0.259	0.00		N											
S6051DA12	2800.000	0.000	0.00		N											
	-821.278	0.966	1.16	1.20	Y											
	864.425	0.259	0.31		N											
S6051DA13	2000.000	0.000	0.00		N											
	-821.021	0.966	0.25	0.25	Y							Y				
	863.464	0.259	0.00		N											
S6051DA14	1842.393	0.000	0.00		N											
	-824.000	1.000	1.20	1.20	Y											
	1186.619	0.000	0.00		N											
M6269M03	3095.484	0.000	2.00	2.00	Y	6.78	0.27									
	-894.685	1.000	1.70	1.70	N											
	1261.374	-0.009	2.00	2.00	Y											
H6409DB01	1911.000	0.000	1.20	2.00	Y	19.00	0.10									
	-801.450	1.000	3.00	3.00	N											
	1623.000	0.000	1.20	1.20	Y											
L6215DC01	3692.328	0.000	1.20	1.20	N			19.00	0.10	25.00	0.10					
	-742.565	1.000	3.00	3.00	N											
	2128.412	0.000	1.20	1.20	Y											
S6031DA03	3357.000	0.000	0.00		N											
	-745.000	0.000	0.00		N											
	1077.000	1.000	0.25	0.25	Y							Y				

Figure 34

Figure 34, above, is a partial example of an aperture assembly measurement point chart. Here, a graphic provides the approximate locations of the points while the chart below lists the specific point nominals, tolerances, whether the point is an alignment point, audit point, or used for a vision system measurement. This chart also represents a breakdown of the tolerance based on the point vector (I,j,k) or the angle of the surface the point is on relative to the main product coordinate system. While most surfaces on this aperture are normal to body grid, in figure x below, provides an example of how the vector is determined.

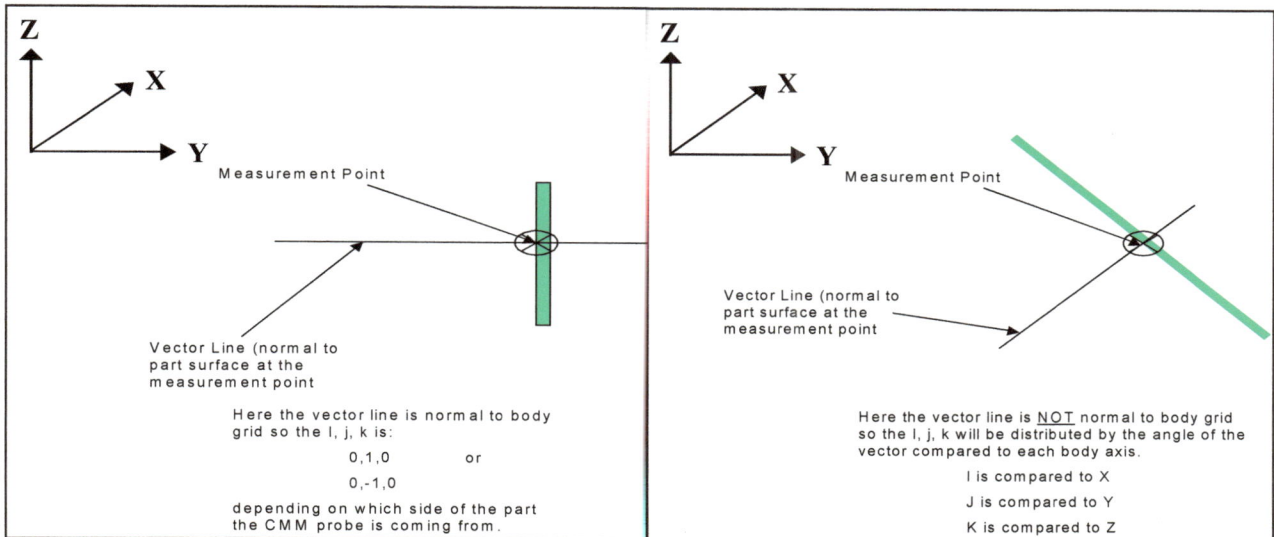

Figure 35 – showing vector examples

Note that in each of the views in figure 35 above, the part (green in color) is infinite along the "X" axis. Having stated this, the I for either case would be "0.000". This information will be used when programming a measurement system so that the system will measure normal to the surface. Thus, providing a more accurate measurement result. Why? This is because an approach vector normal to the surface has eliminated the triangulation error (shown in figure 36 below) that would be present when the part is not perfectly normal. Utilizing the tolerance breakdown for each direction also helps in determining root cause of any build issues.

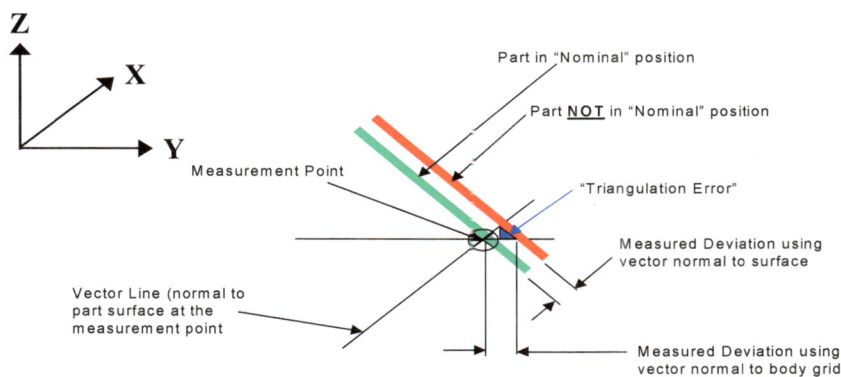

Figure 36

Another example of a measurement chart, below in figure 37, shows points for a vision system. Note: In this section, the nominals are not the same as the nominals in the left hand side of the sheet. Why? This effect is due to what side of material the measurement system will read from. While CMM's have the ability to add material thickness to a nominal provided and calculate the "new nominal" along the vector, vision systems may need this calculated for them, depending on the alignment method used in setting up the vision system.

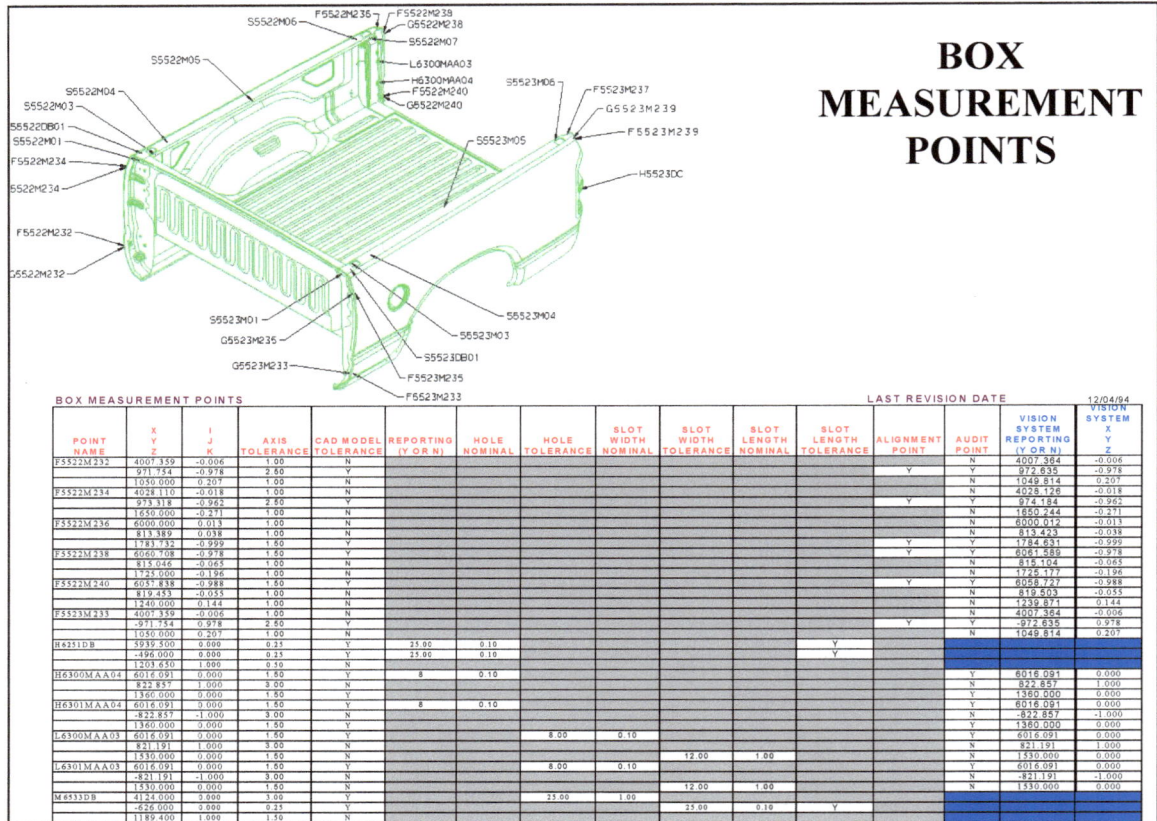

BOX MEASUREMENT POINTS

LAST REVISION DATE 12/04/94

POINT NAME	X Y Z	I J K	AXIS TOLERANCE	CAD MODEL TOLERANCE	REPORTING (Y OR N)	HOLE NOMINAL	HOLE TOLERANCE	SLOT WIDTH NOMINAL	SLOT WIDTH TOLERANCE	SLOT LENGTH NOMINAL	SLOT LENGTH TOLERANCE	ALIGNMENT POINT	AUDIT POINT	VISION SYSTEM REPORTING (Y OR N)	VISION SYSTEM X X Y Z
F5522M232	4007.359	-0.006	1.00	N										N	4007.364	-0.006
	971.754	-0.978	2.50	Y									Y	Y	972.635	-0.978
	1050.000	0.207	1.00	N										N	1049.814	0.207
F5522M234	4028.110	-0.018	1.00	N										N	4028.126	-0.018
	973.318	-0.962	2.50	Y									Y	Y	974.184	-0.962
	1650.000	-0.271	1.00	N										N	1650.244	-0.271
F5522M236	6000.013		1.00	N										N	6000.012	-0.013
	813.389	0.038	1.00	N										N	813.423	-0.038
	1783.732	-0.999	1.50	Y									Y	Y	1784.631	-0.999
F5522M238	6060.708	-0.978	1.50	Y									Y	Y	6061.589	-0.978
	815.046	-0.065	1.00	N										N	815.104	-0.065
	1725.000	-0.196	1.00	N										N	1725.177	-0.196
F5522M240	6057.838	-0.988	1.50	Y									Y	Y	6058.727	-0.988
	819.453	-0.055	1.00	N										N	819.503	-0.055
	1240.000	0.144	1.00	N										N	1239.871	0.144
F5523M233	4007.359	-0.006	1.00	N										N	4007.364	-0.006
	-971.754	0.978	2.50	Y									Y	Y	-972.635	0.978
	1050.000	0.207	1.00	N										N	1049.814	0.207
H6251DB	5939.500	0.000	0.25	Y	25.00	0.10						Y			
	-496.000	0.000	0.25	Y	25.00	0.10						Y			
	1203.650	1.000	0.50	N										
H6300MAA04	6016.091	0.000	1.50	Y	8	0.10							Y	6016.091	0.000
	822.857	1.000	3.00	N									N	822.857	1.000
	1360.000	0.000	1.50	N									Y	1360.000	0.000
H6301MAA04	6016.091	0.000	1.50	Y	8	0.10							Y	6016.091	0.000
	-822.857	-1.000	3.00	N									N	-822.857	-1.000
	1360.000	0.000	1.50	Y									Y	1360.000	0.000
L6300MAA03	6016.091	0.000	1.50	Y			8.00	0.10					Y	6016.091	0.000
	821.191	1.000	3.00	N									N	821.191	1.000
	1530.000	0.000	1.50	N					12.00	1.00			Y	1530.000	0.000
L6301MAA03	6016.091	0.000	1.50	Y			8.00	0.10					N	6016.091	0.000
	-821.191	-1.000	3.00	N									N	-821.191	-1.000
	1530.000	0.000	1.50	N					12.00	1.00			N	1530.000	0.000
M6533DB	4124.000	3.000	3.00	Y			25.00	1.00						
	-626.000	-3.000	0.25	N					25.00	0.10		Y		
	1189.400	1.000	1.50	N										

Figure 37

Below is an example of a measurement point naming convention. The naming convention consists of 5 segments that are combined to describe a measurement point. The objective is so that anyone reviewing data from a particular report, CMM, hand held collector, etc., could easily understand what they are reading. This format is similar to an automobile's VIN (vehicle identification number).

The 5 Segments of a Measurement Point Name

1. WHAT TYPE OF POINT IS IT ? ie: Surface, Hole, Slot ...
2. WHAT PART IS THE POINT ON ? 4 digits
3. WHAT ELEMENT IS IT ? ie: DATUM, AUDIT ...
4. ENGINEERING NAME   -- HOLE ID OR DATUM ID --
5. WHAT SEQUENCE NUMBER IS IT ?

POINT TYPES:

E -	EDGE POINT (Non referenced edge point)
F -	FLUSH
G -	GAP H -HOLE, CIRCLE
K -	STUD (External cylinder)
L -	SLOT (Round Ended)
M -	SLOT (Square Ended)
R -	CENTER OF RADIUS, NOTCH, CUTOUT
S -	SURFACE POINT

ELEMENT TYPES:

D -	DATUM
A -	AUDIT POINT
Q -	CLAMPING POINT
M -	MEASUREMENT POINTS
E -	ASSEMBLY PLANT REQUIRED internally
F -	SOURCE PLANT REQUIRED internally
G -	SUPPLIER PLANT REQUIRED internally

REPORTING DIRECTION:

X -	"X" Direction Only
Y -	"Y" Direction Only
Z -	"Z" Direction Only
P-	"True Position" (Vector)
A -	"X, Y, and Z" Directions reported individually
F -	"X and Y" Directions reported individually
U -	"Y and Z" Directions reported individually
D -	"X and Z" Directions reported individually

Fields in a Point Name

First character	Point type
2nd - 5th character	Last 4 digits of part/assembly number
6th character	Element type
7th, 8th, 9th characters	Engineering name (if applicable noting that the name must start with an ALPHA character and must not exceed 4 ALPHA NUMERIC characters), or the sequence number of point.
10th character	Point Reporting Direction

Chapter 4 - Certification – Tools, Gauges, Molds, and Dies

Certification in a production environment, in a sense, is the act of verifying that the object (gauge, tool, etc.) you are measuring meets tolerance requirements supplied. In the automotive environment, three types of certification requiring at least two types of measurement devices are necessary. The types of certification needed are:

- Gauges – both part and assembly
- Assembly tools and fixtures
- Molds and Dies

Each of these certifications has distinct methods and criteria to ensure that the numbers, which will be generated from each of the above devices (gauges, tools, etc.), are reliable to within a given tolerance. Having stated this, tolerances for some devices needing certification are tighter than others and thus require a different certification measurement device to certify. Additionally, the type of certification data needing to be generated might eliminate some systems from consideration due to programming and other issues. For reference purposes only, the table below provides an overview of when to use specific tools in certain situations.

Device to be Certified	Attribute	Typical Tolerances	Certification Measurement System*
Gauge	Pins	+/-0.05mm	CMM
	Blocks	+/-0.05mm	
Assembly Tooling	Pins	+/-0.25mm	Laser Tracker
	Blocks	+/-0.25mm	Photogrametry
			CMM
Mold/Die			Laser Scanner/Tracker/White Light Scanner
	Die Punch	+/-0.25mm	Photogrametry
	Surface	+/-0.25mm	Vision
			CMM

Notes: Typical tolerances are accepted automotive industry tolerances.
Certification Measurement System based on system accuracy and
repeatability abilities from independent studies.
* - Systems listed are for recommendation purposes only based on
independent accuracy and repeatability studies.

As noted in the table above, the CMM (Coordinate Measurement Machine) can certify all of the devices measured. Due to programming issues, timing constraints and portability issues, the CMM may not be the preferred choice of measurement for assembly tools and dies/molds. Additionally, it may be more beneficial to certify molds and dies in a point cloud mode because it will generate data over the entire scanned surface versus at specific points. The point cloud mode also allows the ability to gather points closer together, in very detailed areas, while spreading apart points in areas that are "flat" or without much detail (see figure 43). This scenario allows the opportunity to re-engineer the mold/die model by creating surface from the point cloud data.

Below is a picture of a transfer die operation. In this example, the die generates both the left hand and right hand parts at the same time. This particular die is the restrike die operation (D4).

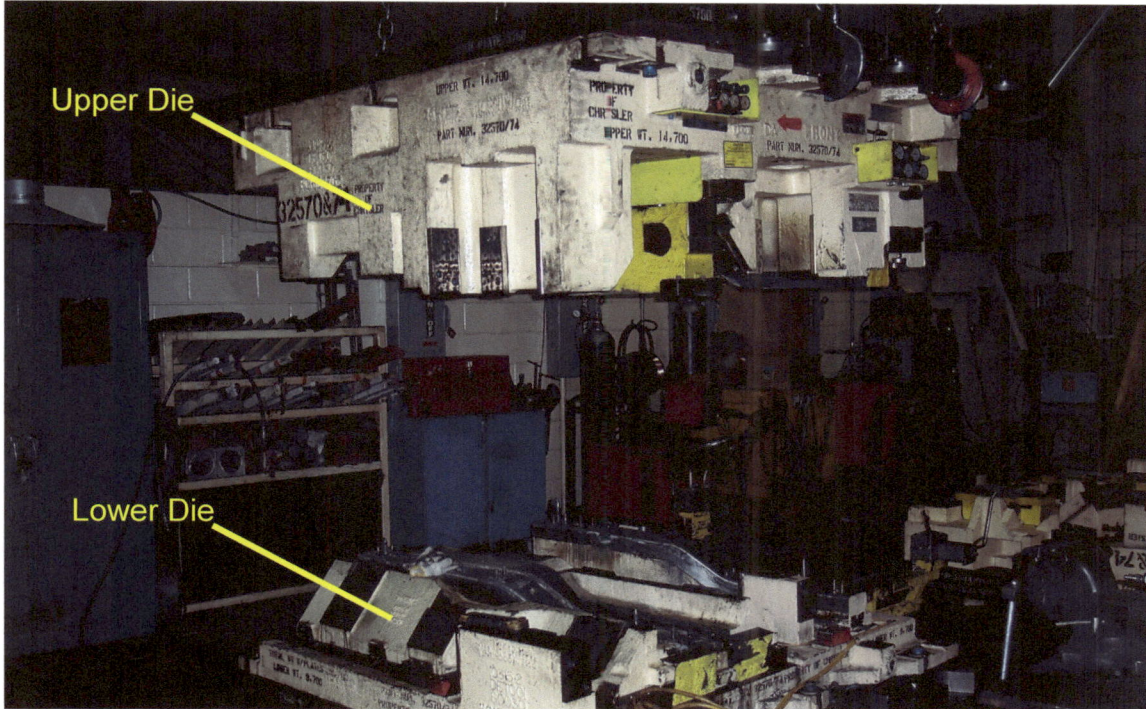

Figure 38

Looking at this die in more detail (figure 40 – upper, figure 39 - lower) shows various wear surfaces, part locators, etc. within the die. Also, noted on the dies are the weights, in this example the lower die weighs 9,700 pounds and the upper die weighs 14,700 pounds and the material flow. The direction of the material flow is necessary so the die is oriented in the die line in the same direction as the other dies in the die line. This is D4 or the restrike die. Typically there are anywhere from four to six dies in a transfer die process. In most cases the die line up is represented by the following:

   D1 – Form Operation
   D2 – Trim Operation
   D3 – Pierce Operation
   D4 – Restrike Operation

Other dies may be added to the die operation for material flow, in case of deep draws and/or spreading out trim or pierce operations due to complexity.

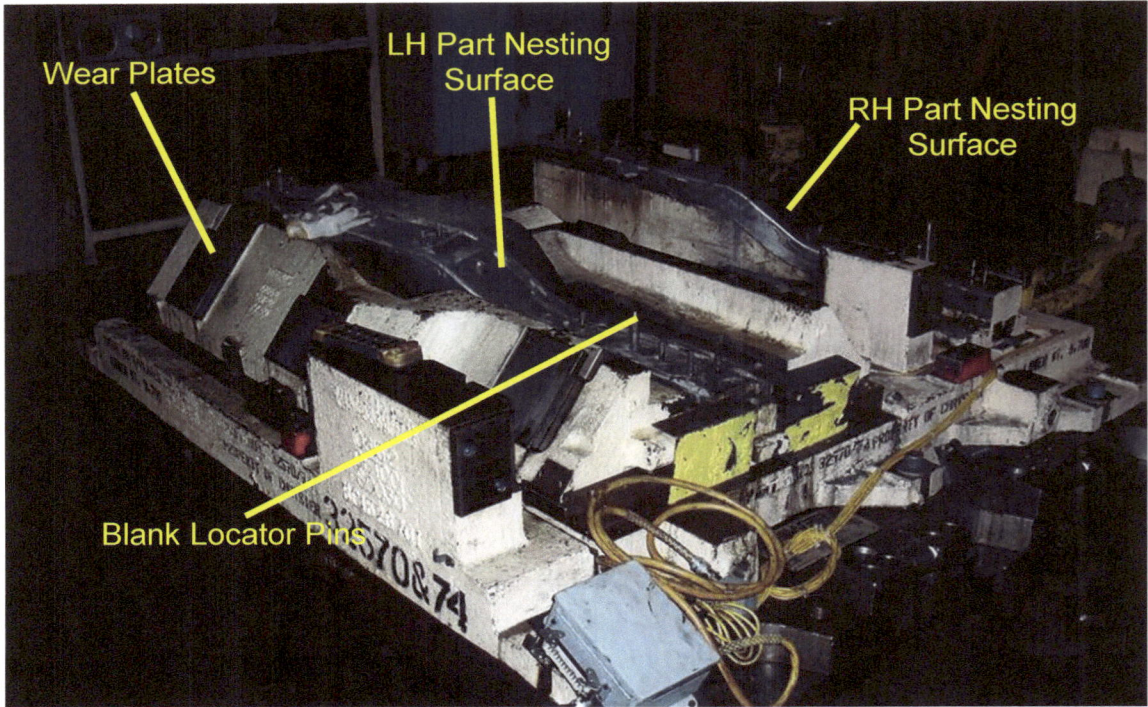

Figure 39 – Lower Die

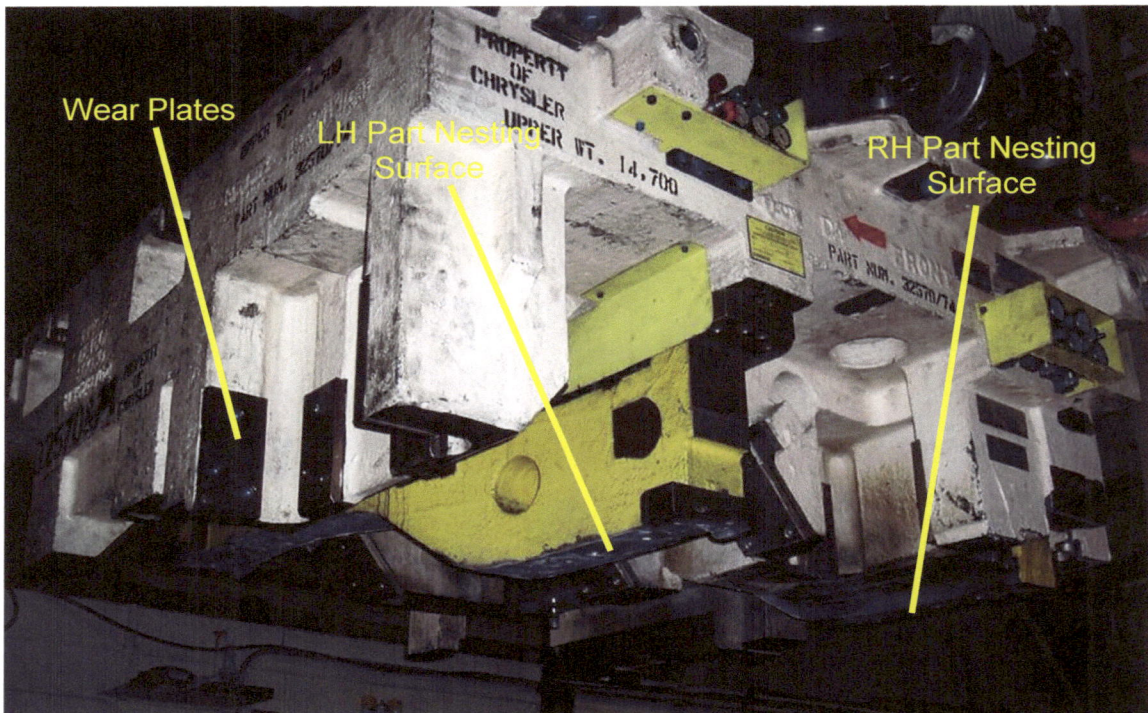

Figure 40 – Upper Die

Using point cloud data for comparison to the product math model also presents a different view. With single point measurement, associated with a CMM, a measured point is compared with a nominal point to produce a deviation. Represented in figure 40, in a point cloud deviation, the cloud is compared to the product model surface and a color graph is generated to show deviation. This color graph can be tailored to show areas with respect to tolerances. It is important to note: while each individual point in a point cloud can generate a specific deviation, due to the amount of points produced in excess of 100,000 in most cases, specific point detail would be enormous and a waste of time.

When using a portable measurement device, it is important to understand two key elements. The calibration results of the system (see figure 41) and the alignment fit. Calibration results indicate how accurately the measurement system a known (certified) value. The alignment fit refers to the datums or benchmarks the system used to get into the part coordinate system. Plus, what directions did the system use for each datum or benchmark (see figure 42). This information will be used by the production run source to verify tooling as part of a maintenance schedule. Any deviation from the original benchmark methods will have an adverse effect on the verification data output.

Calibration Results

```
09/20/01  8:45:42 am
Interim Test Result (After TAC):
Dist(in) AZ(deg) EL(deg) dT(arcsec) dP(arcsec) Tol(in)  Error(in)
-----------------------------------------------------------------
   240     -91      90       1.2         7.7     0.00683   0.00929
   237      -1      90       4.2         6.0     0.00679   0.00845
   240      91      90       1.6         2.3     0.00683   0.00357
    84       1     135       0.8        10.2     0.00425   0.00431

TAC backsight error (After TAC):
Dist(in) AZ(deg) EL(deg) dT(arcsec) dP(arcsec) Tol(in)  Error(in)
-----------------------------------------------------------------
   242    -130      89       0.1         0.7     0.00500   0.00081
   238     -45      90       0.3         1.0     0.00500   0.00119
   238      45      90       0.4         0.1     0.00500   0.00051
   242     130      89       0.3         0.3     0.00500   0.00048
   237       0      90       0.0         0.7     0.00500   0.00081
    81      90     131       0.6         1.6     0.00500   0.00066
    81     -90     131       0.6         1.4     0.00500   0.00061
    79      21     133       0.4         0.2     0.00500   0.00017
    29      91      47       1.1         2.5     0.00500   0.00037
    29     -91      47       4.1         2.2     0.00500   0.00064
    27       1      44       1.3         0.4     0.00500   0.00017
```

Figure 41
The above is an example of a field calibration by a portable measurement system. In this case, a laser tracker was used. This calibration is done to determine if the measurement system is functioning properly by comparing the

measured results to a known value. With some systems, the known value may be a certified invar bar.

```
╔════════════════════════════════════════════════════════════════════════════╗
║              Best-fit to Tooling Ball Results                                ║
╚════════════════════════════════════════════════════════════════════════════╝

SMXInsight 4.00
Job Filename:  2002 DR lh rear quad door outer - Mex - D5        Angle Units:  degrees
Distance Units:  millimeters                                    Date: October-26-2000
Best-fit Frame ...............................................................
Reported in Frame:  @STATION01

PART              Xi  -0.76892441     Xj  -0.63927871     Xk   0.00883051
                  Yi   0.00482743     Yj  -0.01961685     Yk  -0.99979592
                  Zi   0.63932147     Zj  -0.76872486     Zk   0.01816994

                  RX    -88.9588      RY     -0.5060       RZ    -140.2601
                  X    3690.309       Y    3180.687        Z    -1806.720

              Scale   0.99995122

Residuals ....................................................................

Standard Deviation         0.011

tb_01
Actual            X    3634.076       Y    -867.446        Z    1919.987
Nominal           X    3634.079       Y    -867.444        Z    1920.000
Difference        X      -0.003       Y      -0.002        Z      -0.013        0.013

tb_02
Actual            X    3744.733       Y    -956.663        Z    1180.005
Nominal           X    3744.713       Y    -956.664        Z    1180.000
Difference        X       0.020       Y       0.001        Z       0.005        0.021

tb_03
Actual            X    3219.709       Y    -960.374        Z    1180.007
Nominal           X    3219.726       Y    -960.375        Z    1180.000
Difference        X      -0.017       Y       0.001        Z       0.007        0.019

Standard Deviation X       0.019       Y       0.001        Z       0.011
```

Figure 42

Figure 42 is an example of the alignment fit results. In this case, the alignment is based off of three (3) tooling balls. For this report analysis: it is important to understand the scale, the standard deviation of fit, and what directions are used for which alignment features.

A scale equaling a factor of 1.0000 is perfect. Stated another way, the tooling ball measured relationship to each other is exactly the same as the nominal relationship to each other. While it is highly unlikely that a perfect scenario will exist, a number closest to 1.0000 needs to be achieved. Why? So that the scale, of the scanned data, can be compared to the nominal math model with a high degree of confidence. In the above example, the scale factor of 0.99995 was achieved. From the above figure, we can derive that all directions for all the tooling balls were used for the best-fit or alignment results. If any direction or ball were not used, the results would not appear in that column.

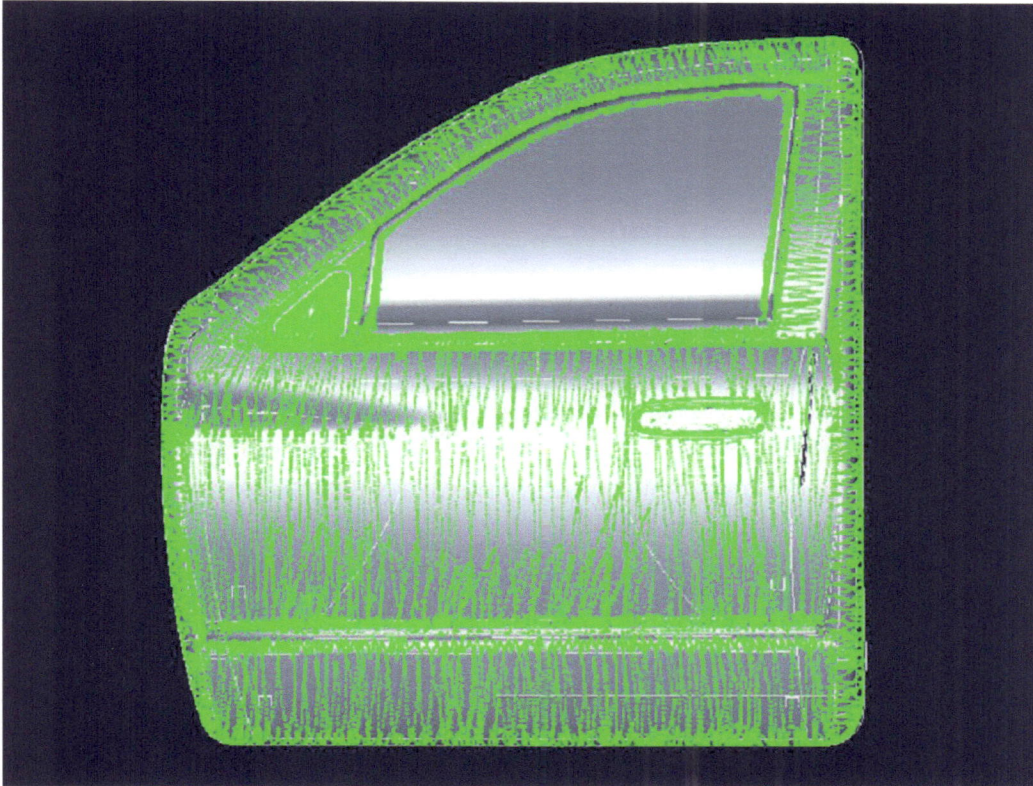

Figure 43 – Point cloud generated from scan (overlayed on development model). Approximately 600,000 points generated.

Once the scanning is completed, a point cloud representing the "as cut die surface" is generated. (See Figure 43.) The data is then compared to the corresponding die model to determine how well the die was cut and relative to design intent (See Figure 45.).

Figure 44

Figure 40 is a laser tracker used for scanning a die. In the larger picture the operator is holding a stylus with a ½" diameter SMR (Spherical Mirrored Retro reflector) that the tracker head follows. This size SMR is typically used for "tight areas" needing detailed data (i.e. the mirror pocket area in figure 43). The larger (1 ½" diameter) SMR (in the smaller right hand picture above) is used for larger surfaces, like the large areas of the door in figure 43.

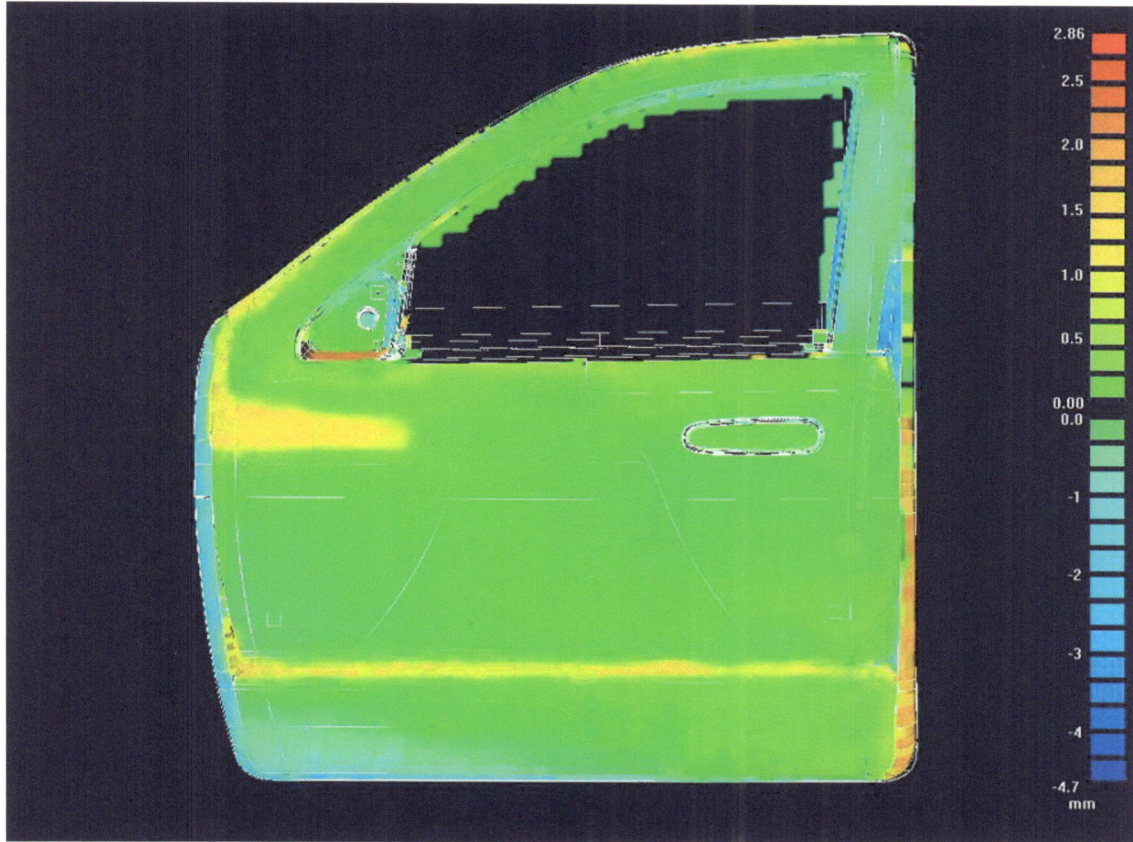

Figure 45 – Scan data to die cut model comparison

At this point, if no major discrepancies exist, the information is documented and the die tryout process begins. If major discrepancies are identified, then the part buyoff team reviews the information and determines the best course of action. Potential outcomes might include, but not be limited to, re-cutting the die or determining initial part quality prior to full die tryout. In this instance, the a-pillar area was re-cut to get the product closer to the product nominal.

After die tryout, the die is rescanned, as represented in figure 46. The diagram depicts greater detail with particular attention to areas where the part, as produced, has significant deviations with respect to the product model. This detail is necessary to capture the needed information to regenerate surfaces and cutter paths for future die/product development. As we can see, Figure 47 highlights these differences.

Once the die surface data is generated, it is then cut into sections at approximately every 100mm apart in body grid. Additional sections are created radially in areas where large radii exist.

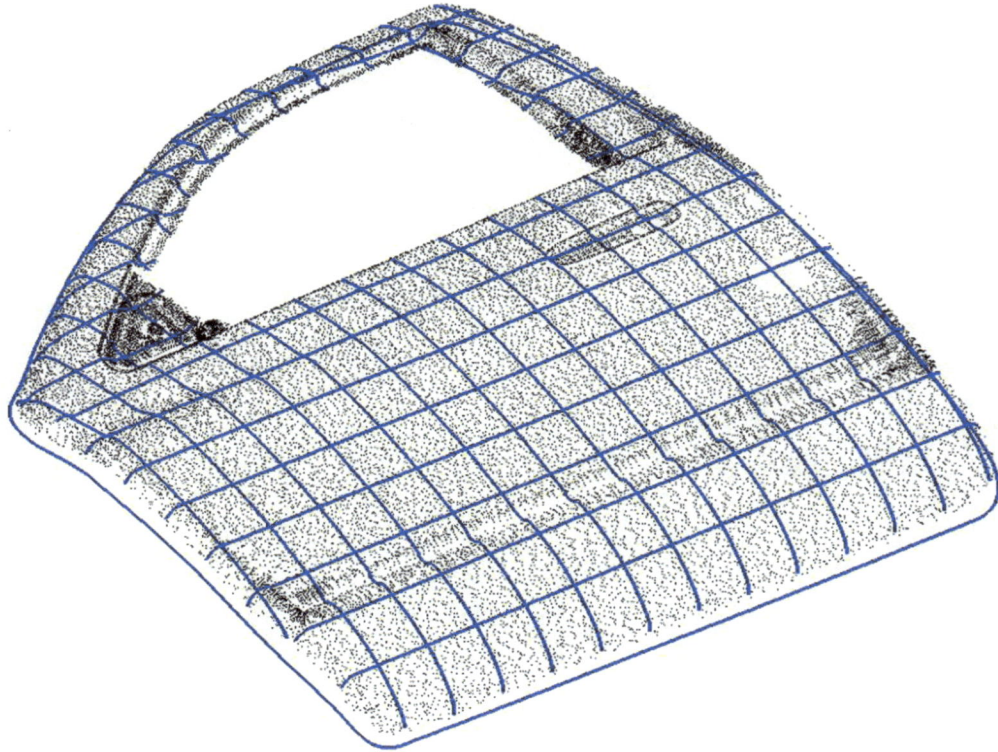

Figure 46 – Point cloud section with generated curves.

The section data, shown in Figure 46, is then used to recreate surface. Thus, creating a new development model and new cutter paths. Comparing the section data, to the original development model, can reduce the time in regenerating new surface where the scan data shows the die has not changed.

Depending on the software used to "reverse engineer" surfaces, the sections may not be necessary as some softwares provide quick surface generation from the point cloud itself.

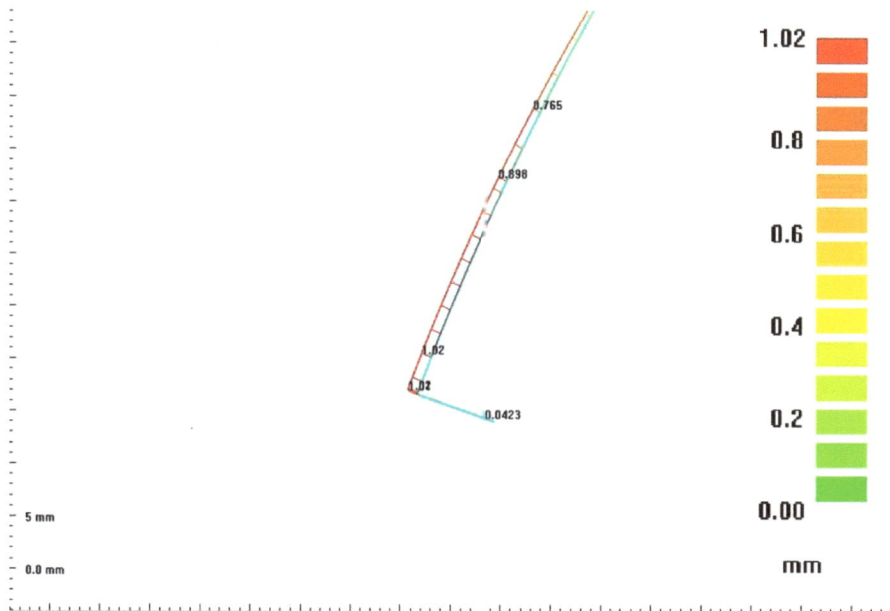

Figure 47 – Detailed view of one section depicting difference between development model and actual scan

The sections generated from the scan data can then be used to re-engineer the part by creating surfaces through the scan data. Once this is completed, the new development model surface is given to the production run source to be used as part of a preventative maintenance program. This new development model should also be given to the die/mold development group. Why? So that changes to the development modeling software or a working library can be developed to shorten future product buy-off times.

Another method of scanning is using a white light scanner. This system (an example shown in figure 48). Using this method, nothing touches the artifact being measured, thus eliminating any deflection possibilities if the artifact isn't rigid.

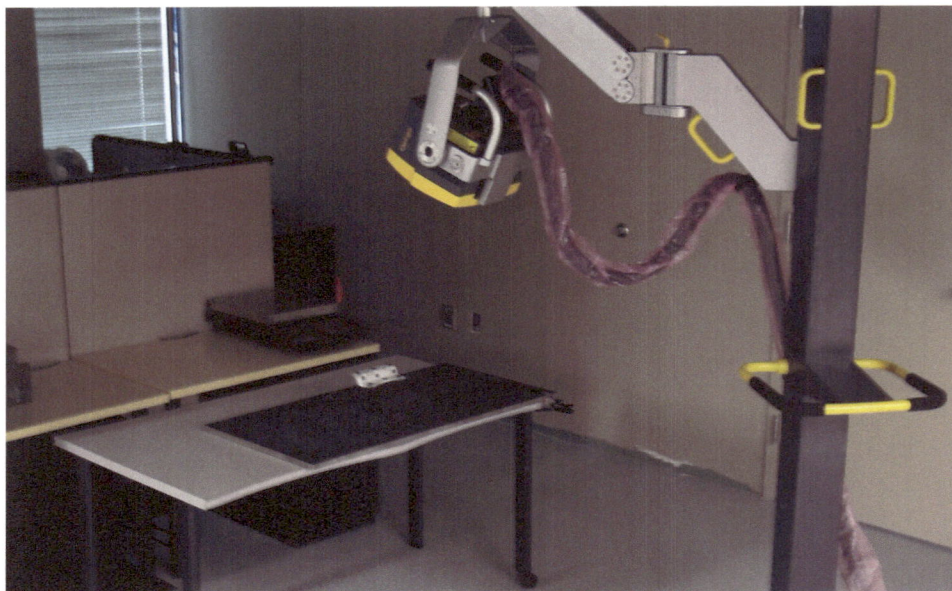

Figure 48 – Example of Portable White Light Scanning System

Depending on the size of the artifact to be scanned, targets are placed on the artifact to "stitch" multiple pictures together. In figure 44, the field of view for this example of a white light scanning system is approx. 12 inches x 18 inches, meaning that artifacts larger than this need to have targets that can be used to stitch pictures together.

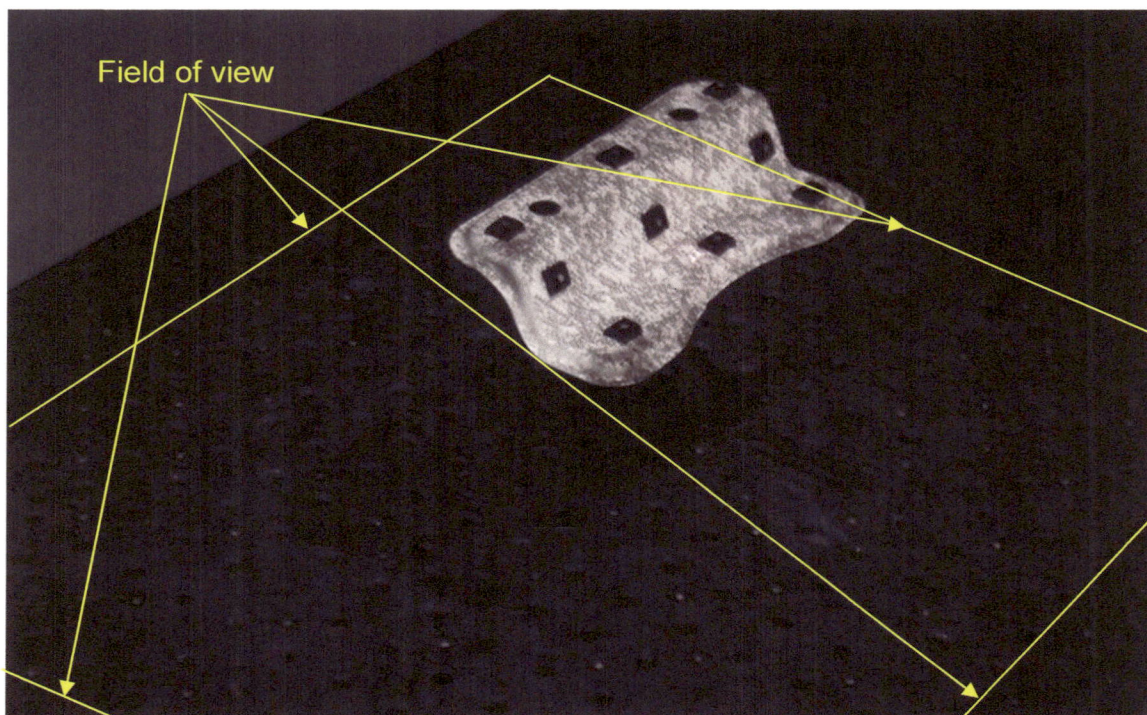

Figure 49 – Field cf view example

Typically, the white light scanner shows a "grid" of the field of view prior to capturing data, this is also accompanied by two laser points. The laser points (shown in figure 50) are a guide for the operator to "focus" the system. As the two laser points converge, the data captured will meet the system's acceptability criteria.

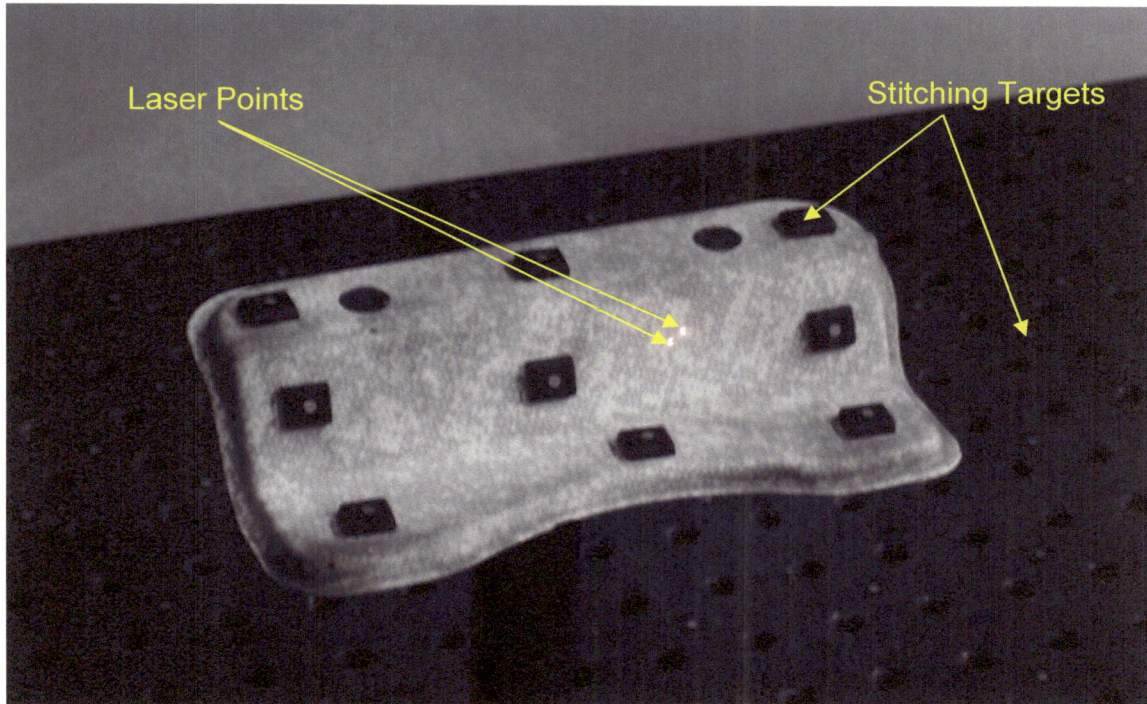

Figure 50 – Laser points cf the white light scanner

In figure 50, stitching targets are placed on the artifact (these are magnetic). Also, in this figure are other stitching target stickers that are placed on a base for multiple part setups. Also note on the artifact, the "grid" light produced by the white light system prior to capturing the image. Typically, white light systems require at least 6 targets within their field of view (4 from the previous scan captured image) to "stitch" images with.

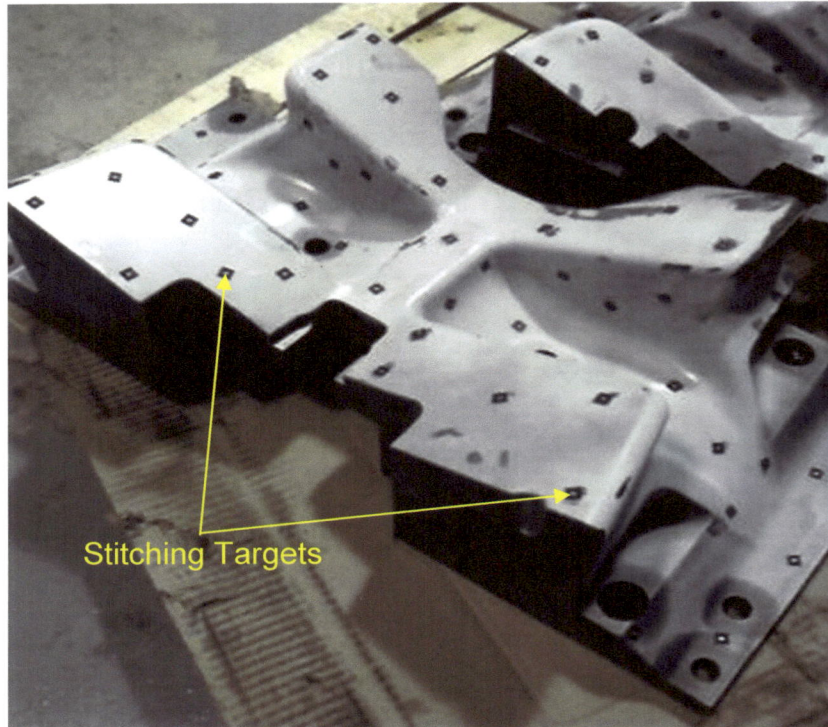

Figure 51 – example die insert

The die insert, in figure 51, is one piece of a particular die. This die was coated with a talc powder to reduce the steel reflection for scanning purposes. As you can see there are numerous "stitch" targets placed on the die insert for scanning purposes. Both the targets and the talc powder will be removed prior to the die insert being installed into the die.

The resulting "color" map (figure 52) compares the die insert to the CAD model and highlights key areas, via annotations, that the die is away from CAD nominal.

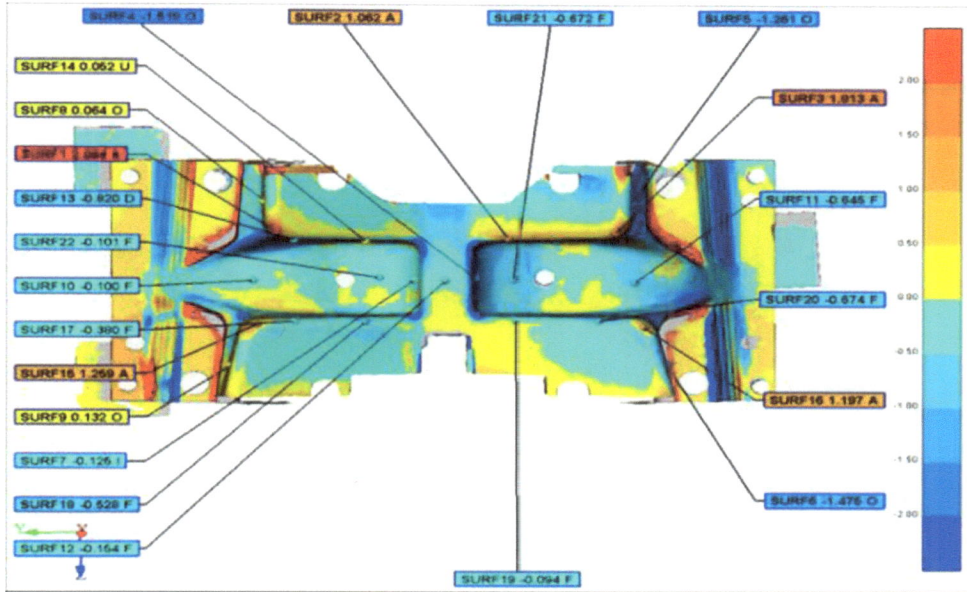

Figure 52 – Die Insert Color Map with Annotations

Another non-contact measurement device is a laser scanner. Pictured below (figure 53), an operator moves the laser head across the part or artifact to gather data. The scanner notifies the operator if the head is too close or too far away to gather data, keeping the data reliable.

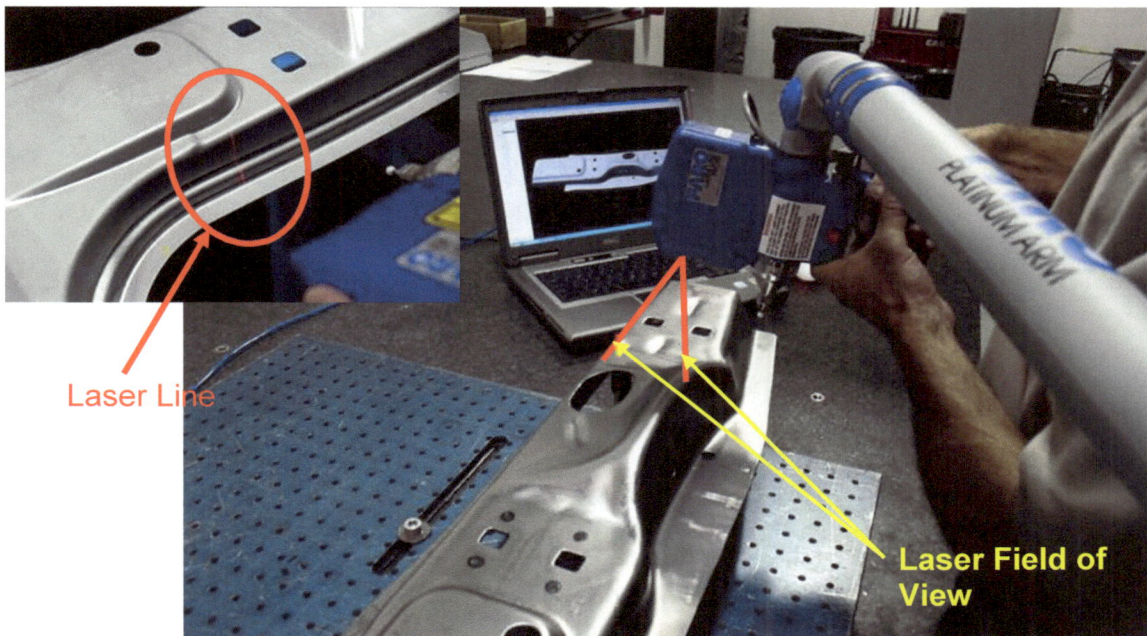

Figure 53

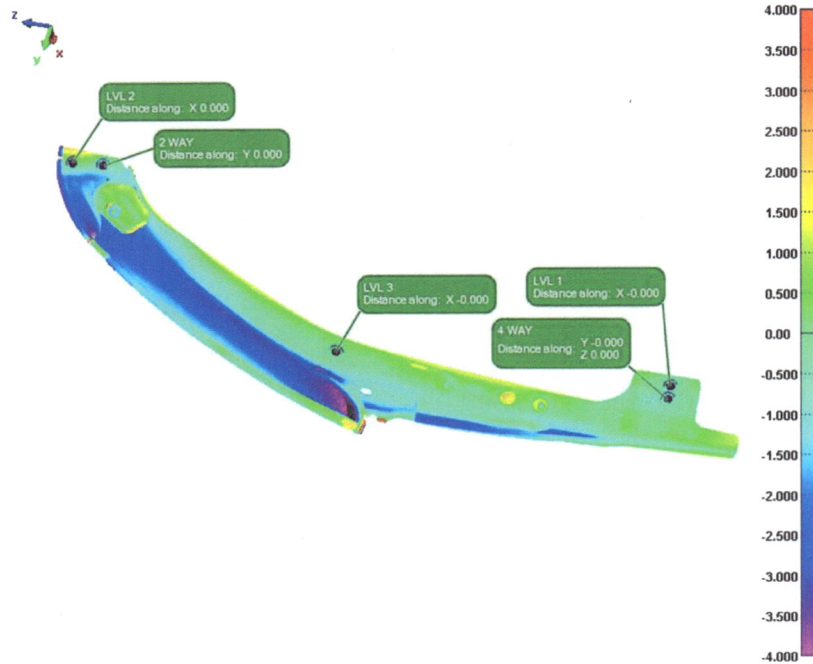

Figure 54 – Laser Scan alignment output

In figure 54, the alignment points used to generate the color map, are identified and show the directions they control along with how well they fit to the CAD, in this case showing 0.000mm deviations.

Points: LVL 1, LVL 2, and LVL 3 are used to create a plane in the X direction
         4-WAY is a locator hole used to locate the part in the Y and Z directions.
         2-WAY is a locator used to locate the part in the Y direction only. These locators create a basic 3-2-1 alignment. If more alignment points are used, the system will create a "best" fit plane through all the points, and then identify a "deviation" at each of these points compared to the "best fit" plane.

No matter what method of measurement is used, it is extremely important to understand the alignment method used to measure the part and generate the data. In the following example, you will see in the various color maps, the difference between the output data based on the alignment. While these examples depict parts, the same alignment considerations for dies, molds and tooling apply.

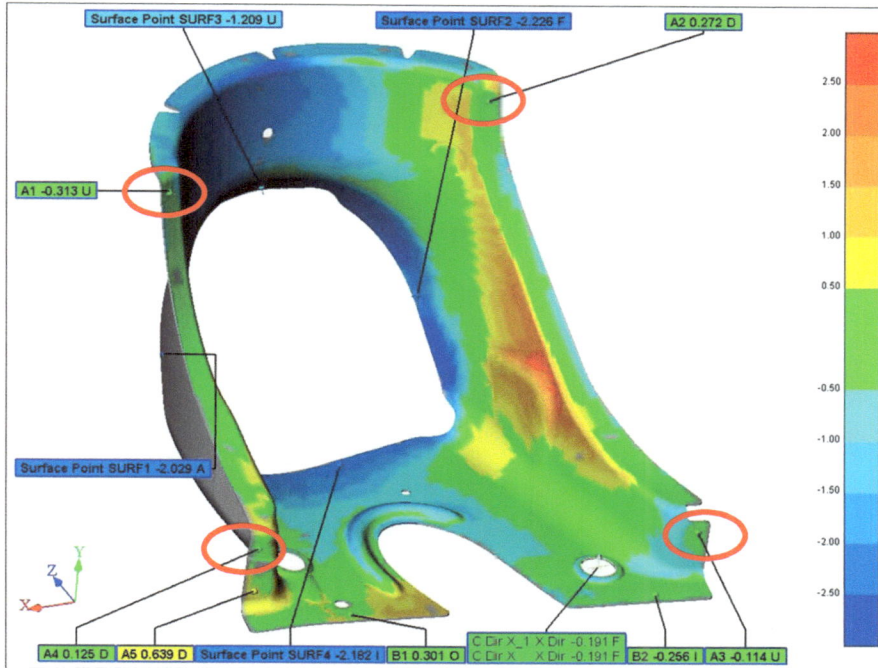

Figure 55 – Alignment using A1 through A4 (circled in RED)

In figure 55, four "A" datums are used to generate the "Z" plane. In this example, the point identified as "A5" has a deviation of 0.639mm (yellow in the color chart).

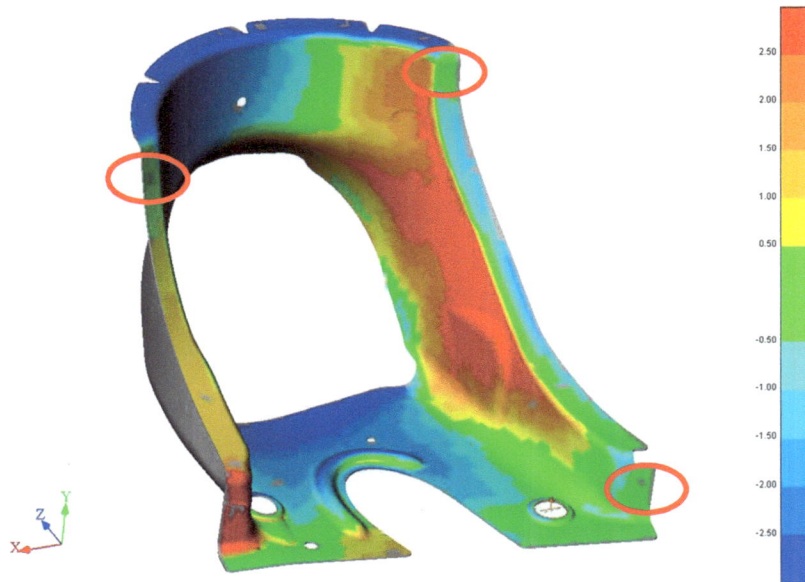

Figure 56 – Alignment using A1, A2 and A3 datums (circled in RED)

Here, in figure 56, with three "A" datums making up the "Z" plane, the resulting data is much different. Here the "A5" point (not shown as an annotation) is RED in the color chart, meaning the deviation is larger than 2.00mm. These two color maps are of the same scanned part, but by changing the alignment method, the resulting output is extremely different.

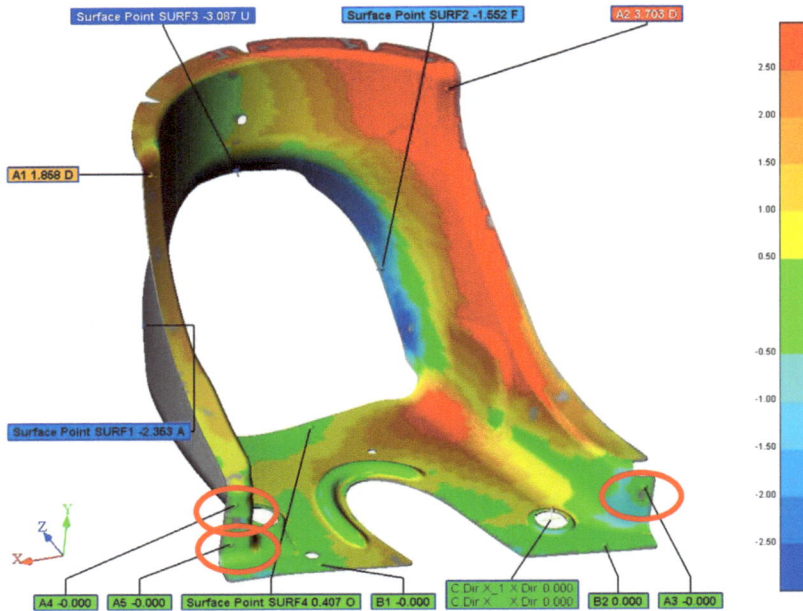

Figure 57 – Alignment using A3 through A5 datums only

Taking this one step further, if we used only the A3 through A5 datums, as in figure 57, the deviations at A1 and A2 are greater than 1.5mm. While this datum scheme is not correct for this part, the point is, that unless you understand the alignment method, the resulting data can have a huge effect on how or where the part will be corrected in the die or assembly process.

Another example of this is in the figures below, here is a 2-out form from the 1st die operation. Here, while we scanned the parts as one, we compared the LH part scan to the LH part CAD, aligning with the LH part datums only. Notice that the RH part is predominately "BLUE" or greater than 2.5mm from nominal. Without understanding the alignment, the right hand part die might be modified where it might not need it.

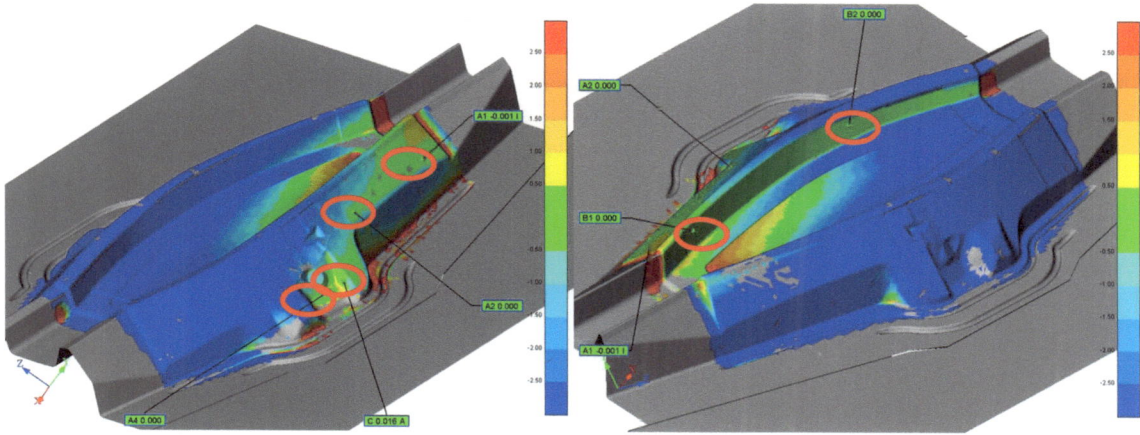

Figure 58 – Scan comparison of 1st operation part to part CAD using left hand part as alignment only.

It is also important to understand what file is being used to compare the scan to. In figure 59, the 1st operation die post (both the left hand and right hand parts are stamped at the same time) is scanned, but instead of comparing it to the product CAD, it is compared to the CAD generated from the 1st operation part scan. This was necessary to determine if the part coming off the die replicated the die itself or if there was any "springback" present. Again, in these color maps, we have aligned only to the LH part

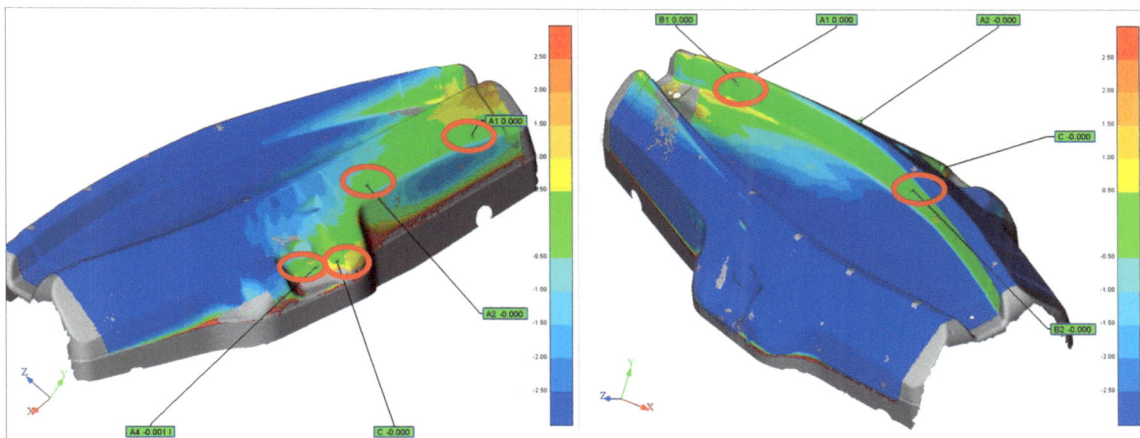

Figure 59 – Die Post scan compared to 1st Operation Part CAD from scan

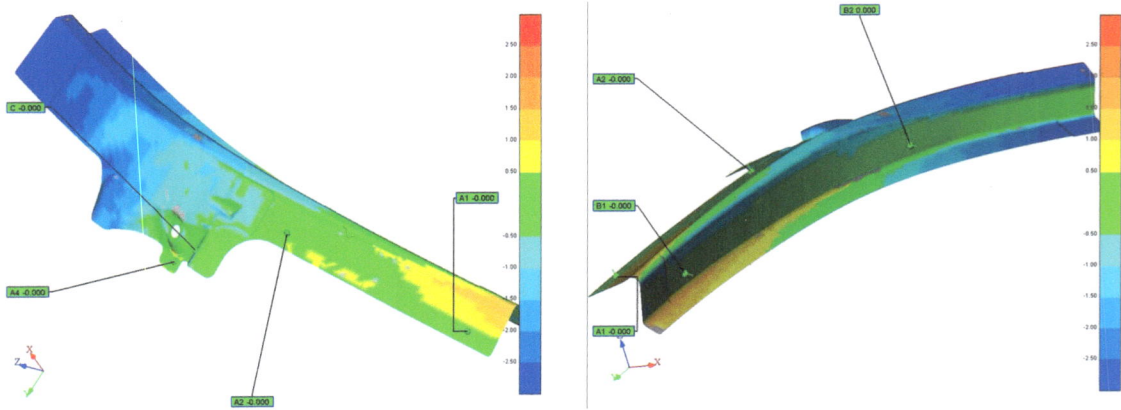

Figure 60 – Left Hand 1st Operation part scan compared to part CAD

Finally, in figure 60, is a left hand part only 1st operation scan compared to the product CAD, using the left hand part datums for alignment. Here the right hand scan data is removed to eliminate any confusion in the resulting data. This is also repeated separately for the right hand part.

This is also true for other measurement devices where the alignment is integral to the measurement. For example, in a CMM measurement, does the CMM align to the fixture benchmarks or to the part itself? If to the part itself, does the program utilize all the datums in the part locating scheme? If there are features of size (holes, slots, etc.), does the GD&T define them as RFS (regardless of feature size), MMC (Maximum Material Condition), or LMC (Least Material Condition) and how is that captured in the program?

This is important as an RFS locator requires the measurement device to merely find the center of the feature, but a MMC or LMC locator not only looks for the center, but depending on the actual hole size of the part, may provide a "bonus" tolerance (size tolerance on the locator that will effect the measurements of other points) that needs to be captured. For more on this topic, refer to Chapter 2, pages 26 and 27.

Assembly Tool Certification

There are a few ways of certifying an assembly tool, either taking specific measurement points on each net block (part locator block) and pin, similar to how a CMM would certify a check fixture, or to scan the surface of the net block and pin to generate the pin center.

In figure 61, an operator, using a laser tracker, holds the SMR against the net surface to capture its location. This step is repeated for all net locations and pin locations.

Figure 61 – Assembly Tool certification using laser tracker

An example of a typical certification measurement output for gauge or assembly tool is provided below.

Example 1 – Gauge/Tool Certification output

Cover Sheet

Road Map of Measured Points

Measurement Output

Chapter 5 – Developing Repeatability Guidelines

Once tolerances have been defined for all parts and processes, two types of repeatability, tool and gauge, are measured and discussed below.

Tool repeatability is performed in two stages, static and dynamic. These repeatability studies determine how well the part can locate consistently in a geometry setting process. Target placement for this study should include the following:

- (1) Target on building floor
- (1) Target on process equipment base
- Minimum of 3 targets on each part in the process. This will help the analyst determine root cause, if an issue arises by eliminating the rotation point that may be present if only 2 targets are provided.

Targets should be placed in area where work is being done in the process equipment but away from any locators in the equipment. Stated another way, place targets where the results will indicate what the process may be doing in the area the parts are assembled; but do not influence the results by placing targets near locators. If locators are set up correctly, then your results will not show any issues at the processing area itself.

The purpose of the targets on the building floor and on the process equipment base is to make sure that nothing gets bumped or moved during the study. The one on the building floor verifies that the measurement equipment stays at the same location and monitors vibration. The one on the process equipment or tool base is to make sure the tooling doesn't move relative to the measurement equipment.

Figure 62 – Showing Laser Tracker tooling repeatability setup

Red Lines identify target locations with respect to the laser tracker itself. Below is a general description of each stage of the tool repeatability study:

- *Static stage* - a part is located in the process and measured 30 times without touching the part or cycling the processing equipment. This study defines the repeatability of the measurement equipment itself. Generally, an acceptable target for this stage is 30% of the process equipment tolerance.
- *Dynamic stage* - a single part measured thirty times, however, the part is loaded and unloaded and the processing equipment is cycled between each measurement. Here, the generally acceptable target is 0.25mm range.

Below is an example of a particular tool study showing placement of the measurement targets and the results of each stage.

TOOL NUMBER: 4432
Measurement Target Locations

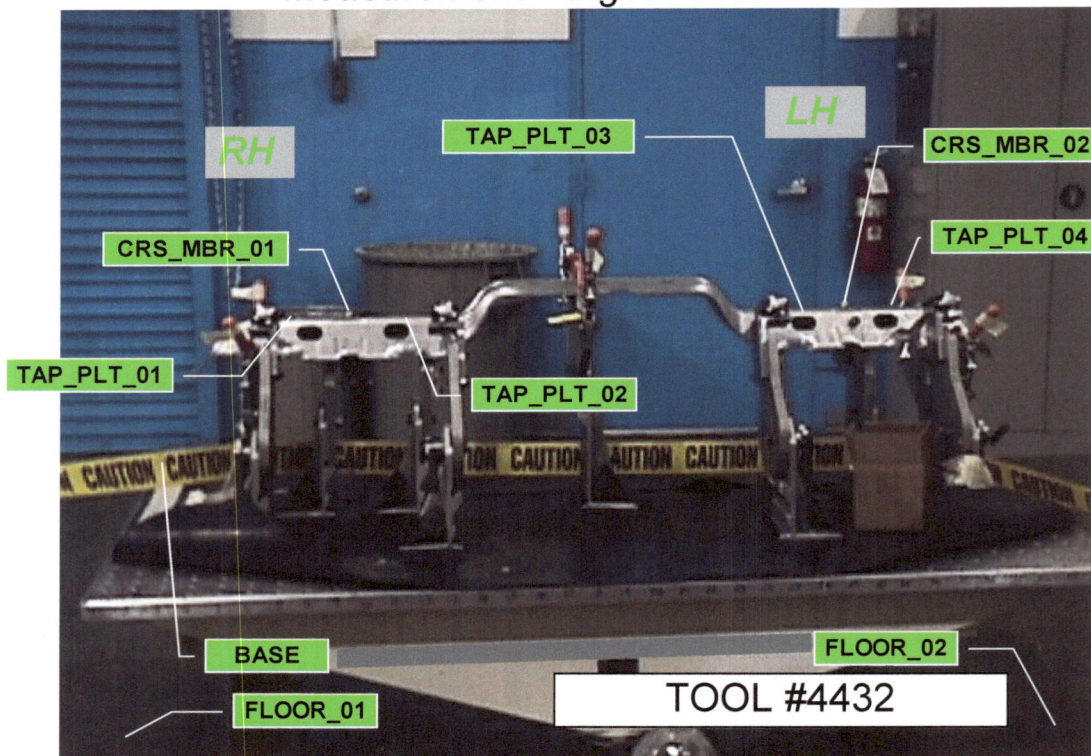

Figure 63 – Tool Repeatability Measurement Target Locations

The figure above is a typical roadmap depicting locations of measurement targets used in the tool repeatability study. Note that in this measurement two targets are placed on the shop floor to detect any vibration, one on the fixture base to detect fixture movement. In this test, 2 targets per part were used due to the access to the parts (tap plates are under the cross member).

STATIC REPEATABILITY RESULTS

Point Name	Six Sigma	Range	Tolerance
BASE(dx)	0.06	0.03	0.25
FLOOR_02(dz)	0.04	0.03	0.25
FLOOR_02(dx)	0.03	0.02	0.25
8826341_LH(dx)	0.03	0.02	0.25
BASE(dz)	0.03	0.02	0.25
8826521_LH_OTR(dx)	0.03	0.01	0.25
8826341_LH(dy)	0.02	0.01	0.25
8826341_RH(dx)	0.02	0.01	0.25
8826521_RH_INR(dx)	0.02	0.01	0.25
8826521_RH_OTR(dx)	0.02	0.01	0.25
8826521_LH_INR(dy)	0.02	0.02	0.25
8826341_RH(dz)	0.02	0.02	0.25
FLOOR_02(dy)	0.02	0.01	0.25
8826521_LH_INR(dx)	0.02	0.01	0.25
FLOOR_01(dz)	0.02	0.01	0.25
8826521_RH_INR(dy)	0.02	0.01	0.25
8826521_LH_INR(dz)	0.02	0.01	0.25
8826521_LH_OTR(dy)	0.02	0.01	0.25
8826521_LH_OTR(dz)	0.02	0.01	0.25
8826341_LH(dz)	0.02	0.01	0.25
FLOOR_01(dx)	0.02	0.01	0.25
BASE(dy)	0.02	0.01	0.25
8826521_RH_OTR(dz)	0.01	0.01	0.25
8826341_RH(dy)	0.01	0.01	0.25
FLOOR_01(dy)	0.01	0.01	0.25
8826521_RH_OTR(dy)	0.01	0.01	0.25
8826521_RH_INR(dz)	0.01	0.01	0.25

Points with 6 Sigma Less Than .25   100

Points with 6 Sigma Greater Than .25 & Less Than .5   0

Points with 6 Sigma Greater Than .5   0

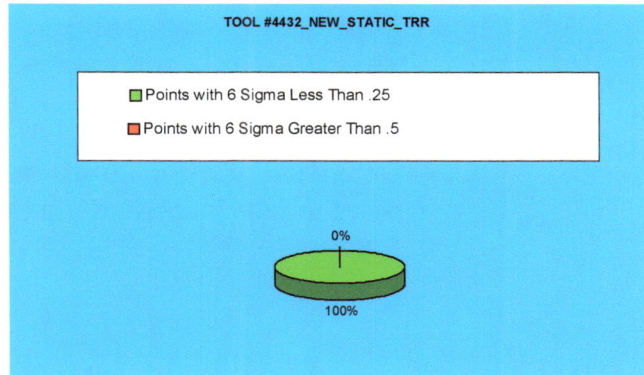

Figure 64 – Static Repeatability Results

Figure 64, shows the results of the static repeatability test ranked from most to least 6-sigma variation. Each direction (dx, dy and dz) for each measurement location is reported. Though some may not have an influence in the overall tool functionality.

DYNAMIC REPEATABILITY RESULTS

Point Name	Six Sigma	Range	Tolerance
8826521_LH_OTR(dx)	0.29	0.14	0.25
8826521_RH_INR(dx)	0.29	0.18	0.25
8826521_LH_INR(dx)	0.22	0.13	0.25
8826341_RH(dx)	0.17	0.09	0.25
8826521_RH_OTR(dx)	0.13	0.06	0.25
8826341_RH(dz)	0.12	0.05	0.25
8826341_LH(dy)	0.12	0.07	0.25
8826341_RH(dy)	0.09	0.05	0.25
8826521_RH_OTR(dy)	0.08	0.04	0.25
8826521_LH_INR(dz)	0.07	0.04	0.25
8826521_LH_INR(dy)	0.07	0.04	0.25
8826341_LH(dx)	0.07	0.03	0.25
8826521_RH_INR(dy)	0.06	0.03	0.25
8826521_RH_INR(dz)	0.06	0.03	0.25
8826521_LH_OTR(dy)	0.06	0.03	0.25
8826521_LH_OTR(dz)	0.04	0.02	0.25
8826521_RH_OTR(dz)	0.04	0.02	0.25
8826341_LH(dz)	0.03	0.02	0.25
FLOOR_02(dx)	0.03	0.02	0.25
FLOOR_01(dz)	0.03	0.01	0.25
FLOOR_01(dx)	0.03	0.01	0.25
FLOOR_02(dz)	0.03	0.02	0.25
BASE(dx)	0.02	0.01	0.25
FLOOR_02(dy)	0.02	0.01	0.25
FLOOR_01(dy)	0.02	0.01	0.25
BASE(dz)	0.02	0.01	0.25
BASE(dy)	0.01	0.01	0.25

Points with 6 Sigma Less Than .25   92.59259

Points with 6 Sigma Greater Than .25 & Less Than .5   7.407407

Points with 6 Sigma Greater Than .5   0

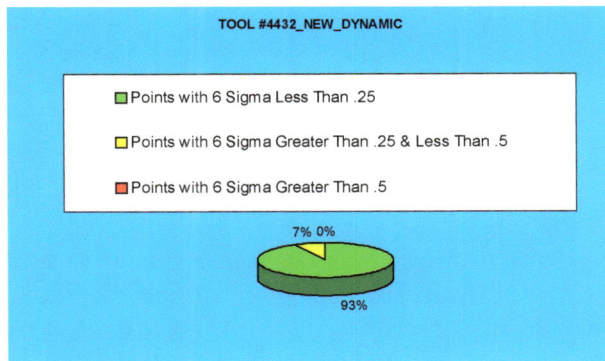

Figure 65 – Dynamic Repeatability Results

In figure 65 above, the dynamic repeatability results are reported exactly like that of the static repeatability results. This figure, however, shows a difference from the static results and dynamic results on the same tool.

Gauge repeatability is performed in two ways: GR (Gage Repeatability) and a GR&R (Gauge Repeatability and Reproducibility) studies. In each case, the part(s) are loaded into the gauge according to the gauge sequence of operations and measured. The process is then repeated, with the amount of times, based on the type of study performed. The differences between the two gauge studies are the following:

- *Gauge Repeatability* – requires only a single part that is measured ten times. This generates a percentage of part tolerance used by the gauge itself. Generally speaking, a GR should be performed in the following scenarios:
  - An initial study when less than five parts are available
  - When the part is located via the measurement system (where the measurement system aligns to the part itself)

GR Procedure: (1-1-10) - 1-part, 1-operator, 10-trials

1. Fixture must be certified
2. 1 part must be selected
3. 1 operator will load and clamp the part following clamp sequence in outlined in the product math model
4. All measurement points must be measured, unless team consensus is achieved on a subset.
5. Repeat steps 3 and 4 until 10 samples are measured
6. Calculate Gage R from all 10 samples
7. Summarize Gage R study
   - Graphic showing point locations
   - Table showing point name and GR %

Gauge Repeatability (GR) Percentage Criteria is as follows:

Under 10% error	- Gage system O.K.
10% to 20% error	- May be acceptable based upon importance of application, cost of gage, cost of repairs, etc.
Over 20% error	- Gauge system needs improvement. Make every effort to identify the problems and have them corrected.

More error requires a review of the gauge and locating scheme, in the failing area, to determine a course of action. This may include the addition of locators, a change in part clamping sequence, or a tolerance concession.

The GR percentage criteria are based in part on the GR&R percentage criteria outlined in the Measurement Systems Analysis Manual. Previous experience has shown that a GR meeting a maximum 20% error can pass a GR&R with less than 30% error. Additionally, a GR with greater than 20% error will not pass

GR&R requirements at the same point. An example of a GR report, with a completed form, is shown in figure 66 denoted on the next page.

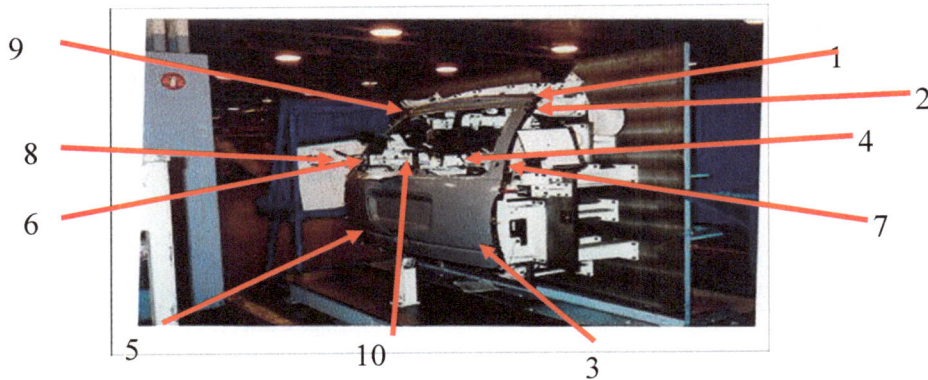

PART/SUBASSEMBLY NAME :   Liftgate Assembly   Pt. #45332376AA

TRIALS	MEASUREMENT POINT LOCATIONS									
	1	2	3	4	5	6	7	8	9	10
1	0.26	-0.05	0.55	1.02	0.33	0.40	0.10	0.22	0.22	0.70
2	0.30	0.02	0.62	1.01	0.45	0.38	0.12	0.27	0.24	0.78
3	0.26	0.10	0.67	1.05	0.49	0.32	0.10	0.42	0.28	0.65
4	0.22	0.03	0.80	1.03	0.25	0.27	0.09	0.23	0.29	0.82
5	0.34	-0.03	0.72	1.10	0.21	0.33	0.16	0.35	0.11	0.90
6	0.28	-0.02	0.65	0.98	0.49	0.40	0.02	0.36	0.14	0.76
7	0.23	-0.04	0.78	0.96	0.22	0.40	0.12	0.27	0.27	0.65
8	0.21	0.02	0.59	1.03	0.48	0.36	0.14	0.39	0.11	0.91
9	0.19	0.10	0.71	1.00	0.53	0.32	0.10	0.21	0.29	0.69
10	0.18	0.03	0.67	0.98	0.20	0.29	0.09	0.28	0.24	0.72
TOL RANGE	1.20	2.00	1.00	2.00	2.00	1.00	2.40	1.50	1.50	1.20
RANGE	0.17	0.15	0.25	0.14	0.33	0.13	0.14	0.21	0.18	0.26
AVERAGE	0.25	0.02	0.68	1.02	0.37	0.35	0.10	0.30	0.22	0.76

	MEASUREMENT POINT LOCATIONS									
	1	2	3	4	5	6	7	8	9	10
GAGE R%	22.8%	14.1%	42.0%	10.7%	35.9%	25.4%	8.2%	26.4%	25.6%	41.7%

$$\text{GAGE R\%} = \frac{\{\text{SQRT}[\text{ SUM}(X - X\text{ Bar}) \ / (n-1)] \times [5.15 / 0.972]\} \times 100}{\text{TOLERANCE RANGE}}$$

Figure 66

Above is an example of a GR study utilizing a manual data collector for a liftgate. Here, 10 points are measured representing all directions of part movement and extreme locations the part will be measured at. While this part may have more than 10 total measurement points, the subset used for the GR provides an indication of how repeatable measurement points near a respective GR point will be.

Another example of a GR output is that from a CMM. Below is a typical GR report from a CMM statistical package. As with any automatically generated results, manual verification of the reported measurement errors should be made at least once per system, in order to ensure that the statistical package used is correct.

GAGE Repeatability Report

PART NAME:													DATE:			
FIXTURE:													PAGE: 1 OF 4			

LABEL	BLK/EL		1	2	3	4	5	6	7	8	9	10	TOLR	RNGE	X-BAR	GAGE R
F6120M228	20	Z	1.56	1.56	1.57	1.59	1.60	1.63	1.65	1.64	1.65	1.64	2.88	0.10	1.61	6.9%
F6120M228	20	P	1.78	1.80	1.79	1.82	1.83	1.36	1.88	1.87	1.88	1.87	3.00	0.10	1.84	7.0%
F6120M228R	30	Y	1.58	1.65	1.57	1.66	1.62	1.54	1.65	1.65	1.64	1.62	0.70	0.09	1.63	23.7%
F6120M228R	30	Z	1.50	1.50	1.51	1.53	1.54	1.57	1.59	1.58	1.59	1.59	2.92	0.09	1.55	7.0%
F6120M228R	30	P	1.72	1.73	1.73	1.76	1.77	1.80	1.82	1.81	1.82	1.81	3.00	0.10	1.78	7.1%
G6120M228	40	Y	1.16	1.24	1.15	1.24	1.20	1.22	1.23	1.23	1.22	1.19	2.89	0.09	1.21	5.8%
G6120M228	40	Z	1.34	1.35	1.35	1.37	1.39	1.41	1.43	1.42	1.43	1.43	0.79	0.09	1.39	25.2%
G6120M228	40	P	1.50	1.52	1.51	1.54	1.55	1.58	1.60	1.59	1.60	1.59	3.00	0.10	1.56	7.0%
F5814M228	50	Z	-0.08	-0.06	-0.07	-0.04	-0.03	-0.01	0.01	0.01	0.03	0.02	2.77	0.10	-0.02	7.8%
F5814M228	50	P	0.13	0.15	0.13	0.17	0.18	0.21	0.22	0.22	0.24	0.23	3.00	0.11	0.19	7.5%
F6120M230	70	Z	1.44	1.45	1.45	1.47	1.48	1.51	1.52	1.52	1.53	1.53	2.90	0.10	1.49	7.0%
F6120M230	70	P	1.58	1.61	1.59	1.63	1.63	1.67	1.69	1.68	1.69	1.68	3.00	0.11	1.64	7.5%
F6120M230R	80	Y	1.02	1.12	1.00	1.11	1.06	1.09	1.13	1.13	1.13	1.09	0.74	0.13	1.09	32.4%
F6120M230R	80	Z	1.25	1.27	1.27	1.29	1.31	1.33	1.35	1.34	1.36	1.35	2.91	0.10	1.31	7.2%
F6120M230R	80	P	1.39	1.42	1.40	1.44	1.45	1.48	1.50	1.50	1.51	1.50	3.00	0.11	1.46	7.7%
G6120M230	90	Y	0.62	0.71	0.60	0.70	0.66	0 68	0.73	0.72	0.72	0.68	2.91	0.13	0.68	8.0%
G6120M230	90	Z	1.18	1.19	1.19	1.21	1.22	1 25	1.27	1.26	1.28	1.27	0.74	0.10	1.23	28.3%
G6120M230	90	P	1.26	1.28	1.26	1.30	1.31	1 34	1.36	1.36	1.36	1.37	3.00	0.11	1.32	7.7%
F5814M230	100	Z	-0.28	-0.27	-0.28	-0.25	-0.24	-0 21	-0.20	-0.20	-0.19	-0.19	2.77	0.10	-0.23	7.1%
F5814M230	100	P	-0.16	-0.13	-0.15	-0.11	-0.11	-0.08	-0.06	-0.07	-0.05	-0.06	3.00	0.11	-0.10	7.3%
F5856M214	120	Z	2.26	2.28	2.29	2.30	2.33	2.36	2.37	2.38	2.39	2.38	2.89	0.13	2.33	8.9%
F5856M214	120	P	2.39	2.42	2.41	2.44	2.46	2.49	2.51	2.53	2.54	2.51	3.00	0.15	2.47	9.3%
F5856M214R	130	Y	0.93	1.05	0.93	1.02	0.98	0.99	1.07	1.07	1.06	1.01	0.75	0.14	1.01	38.0%
F5856M214R	130	Z	2.06	2.08	2.09	2.10	2.13	2.16	2.16	2.18	2.19	2.18	2.90	0.13	2.13	8.7%
F5856M214R	130	P	2.18	2.22	2.21	2.23	2.26	2.29	2.30	2.32	2.33	2.31	3.00	0.15	2.26	9.2%
G5856M214	140	Y	0.54	0.65	0.53	0.62	0.58	0.60	0.68	0.67	0.66	0.62	2.90	0.14	0.61	9.5%
G5856M214	140	Z	2.02	2.05	2.05	2.06	2.09	2.12	2.12	2.14	2.15	2.14	0.75	0.13	2.10	34.2%
G5856M214	140	P	2.08	2.12	2.11	2.13	2.16	2.19	2.20	2.22	2.23	2.21	3.00	0.15	2.17	9.3%
F5814M214	150	Z	0.00	0.01	0.01	0.04	0.05	0.09	0.09	0.09	0.10	0.10	2.76	0.10	0.06	7.2%
F5814M214	150	P	0.10	0.13	0.11	0.15	0.15	0.19	0.20	0.20	0.21	0.20	3.00	0.11	0.16	7.4%
F5856M212	170	Z	2.15	2.16	2.17	2.19	2.22	2.25	2.25	2.27	2.28	2.28	2.74	0.13	2.22	9.9%
F5856M212	170	P	2.55	2.59	2.57	2.61	2.63	2.66	2.68	2.69	2.70	2.69	3.00	0.16	2.64	10.0%
F5856M212R	180	Y	2.04	2.18	2.03	2.15	2.09	2.11	2.19	2.16	2.17	2.10	0.91	0.16	2.12	33.4%
F5856M212R	180	Z	2.07	2.08	2.09	2.11	2.14	2.17	2.17	2.19	2.20	2.19	2.75	0.13	2.14	9.6%
F5856M212R	180	P	2.46	2.50	2.49	2.52	2.55	2.58	2.59	2.61	2.62	2.60	3.00	0.15	2.55	9.8%
G5856M212	190	Y	1.62	1.76	1.62	1.73	1.66	1.68	1.77	1.73	1.74	1.68	2.81	0.15	1.70	10.1%
G5856M212	190	Z	1.94	1.95	1.96	1.98	2.01	2.04	2.04	2.06	2.07	2.06	0.70	0.13	2.01	37.2%
G5856M212	190	P	2.26	2.29	2.28	2.31	2.34	2.37	2.38	2.40	2.41	2.39	3.00	0.15	2.34	9.7%
F5814M212	200	Z	0.22	0.22	0.22	0.25	0.27	0.29	0.30	0.31	0.32	0.32	2.80	0.10	0.27	8.0%
F5814M212	200	P	0.48	0.51	0.49	0.54	0.54	0.57	0.60	0.59	0.61	0.60	3.00	0.12	0.55	8.4%
F6208M204	220	Z	-0.64	-0.58	-0.63	-0.58	-0.60	-0.56	-0.53	-0.56	-0.54	-0.56	1.44	0.11	-0.58	13.0%
F6208M204	220	P	-0.70	-0.64	-0.69	-0.64	-0.66	-0.62	-0.59	-0.62	-0.60	-0.61	3.00	0.11	-0.64	6.6%
G6208M204	230	X	0.64	0.74	0.64	0.69	0.63	0.64	0.68	0.61	0.62	0.58	1.44	0.16	0.65	17.3%
G6208M204	230	Y	-0.06	-0.01	-0.05	-0.02	-0.04	-0.03	-0.01	-0.03	-0.02	-0.04	0.33	0.05	-0.03	26.2%
G6208M204	230	P	-0.87	-0.86	-0.86	-0.83	-0.82	-0.79	-0.77	-0.77	-0.75	-0.76	3.00	0.12	-0.81	7.7%
G5733M204	240	X	-0.18	-0.07	-0.19	-0.13	-0.19	-0.19	-0.15	-0.22	-0.21	-0.25	1.41	0.18	-0.18	19.3%
G5733M204	240	Y	-0.62	-0.56	-0.61	-0.57	-0.61	-0.59	-0.56	-0.60	-0.58	-0.60	0.49	0.07	-0.59	23.9%
G5733M204	240	P	-1.91	-1.85	-1.89	-1.85	-1.87	-1.83	-1.80	-1.83	-1.81	-1.83	3.00	0.12	-1.85	6.5%
F5733M204	250	Z	-1.82	-1.82	-1.81	-1.76	-1.74	-1.74	-1.72	-1.74	-1.73	-1.74	1.44	0.10	-1.76	12.3%
F5733M204	250	P	-1.87	-1.82	-1.86	-1.82	-1.83	-1.79	-1.77	-1.79	-1.77	-1.78	3.00	0.11	-1.81	6.4%
F5856M208	280	Y	0.69	0.86	0.69	0.83	0.76	0.80	0.90	0.82	0.84	0.79	1.50	0.21	0.80	24.8%
F5856M208	280	P	0.68	0.86	0.68	0.83	0.76	0.80	0.91	0.83	0.86	0.80	3.00	0.23	0.80	13.0%
G5856M208	290	X	0.30	0.37	0.28	0.32	0.27	0.27	0.29	0.23	0.23	0.21	1.50	0.16	0.28	16.9%
G5856M208	290	P	0.69	0.86	0.69	0.84	0.77	0.81	0.91	0.85	0.87	0.81	3.00	0.23	0.81	13.2%
G6208M208	300	X	0.83	0.90	0.82	0.86	0.80	0.80	0.82	0.76	0.77	0.74	1.50	0.16	0.81	17.0%

Rev 041699

Figure 67

Figure 67, depicts a typical CMM gauge repeatability report. It contains all the raw data, tolerances, point names, and measurement error similar to the previous example, except in a different format.

Gauge Repeatability and Reproducibility – requires a minimum of five parts measured a minimum of once apiece by two operators, commonly referred to as a 5-2-1. This also generates a percentage of part tolerance used by the gauge but also includes operator influence in measurement. Generally speaking, less than 30% measurement error is deemed acceptable. More error requires a review of the gauge and locating scheme in the area failing to determine course of action. This may include the addition of locators, a change in part clamping sequence or a tolerance concession.

- Other GR&R formats include:
  - 5-2-2 (5 parts measured 2 times by 2 operators each)
  - 10-3-2 (10 parts measured 3 times by 2 operators each)
  - 10-2-3 (10 parts measured 2 times by 3 operators each)
  - 5-1-2 (5 parts measured once by 2 operators each)

Gauge Repeatability and Reproducibility Study Procedure

The following summary identifies requirements for conducting a Gauge Repeatability and Reproducibility Study for all component part and assembly gages.

- A gauge repeatability and reproducibility (GR&R) study is required for all production checking fixtures.
- Tolerance definition is defined by the GD&T found in product CAD (math) model.
- Measurement point definition is as defined within GD&T packages.
- A member of the dimensional control team must be present during GR&R study.
- GR&R should include all critical measurement points. Additional points maybe requested by the approval team.
- GR&R studies are performed using A.I.A.G. standards.
- GR&R must meet passing requirements as stated in the Measurement Systems Analysis Manual.

GR&R Procedure: Using (5-2-1) 5-parts, 2-operators, 1-trial as example

1. Fixture must be certified
2. 5 parts must be selected
3. 2 operators (1 trial each) will load and clamp the parts following clamp sequence outlined in the product math (CAD) model
4. All critical measurement points must be measured
5. Repeat steps 3 and 4 until each operator measures each part one time
6. Calculate Gage R&R
7. Summarize Gage R&R study
   - Graphic showing point locations
   - Table showing point name and GR&R %

Gauge Repeatability & Reproducibility Percentage Criteria is as follows:

Under 10% error	-Gauge system O.K.
10% to 30% error	-May be acceptable based upon importance of application, cost of gage, cost of repairs, etc.
Over 30% error	-Gauge system needs improvement. Make every effort to identify the problems and have them corrected.

GAGE REPEATABILITY & REPRODUCIBILITY STUDY

VENDOR NAME:
PART DISCRIPTION: Quad Aperture
PART NUMBER: 4769322AA
DATE: 3/22/1994
POINT #   M02

COLUMN NO.	1	2	3	4	5	6	7
INSPECTOR	A --			B --			
SAMPLE #	TRIAL #1	TRIAL #2	DIFF.	TRIAL #1	TRIAL #2	DIFF.	X-BARp
1	-0.270	-0.239	0.031	-0.243	-0.236	0.007	-0.247
2	-0.083	-0.047	0.036	-0.063	-0.057	0.006	-0.063
3	-0.334	-0.364	0.030	-0.382	-0.380	0.002	-0.365
4	-0.332	-0.375	0.043	-0.342	-0.338	0.004	-0.347
5	-0.217	-0.210	0.007	-0.215	-0.226	0.011	-0.217
6	-0.270	-0.239	0.031	-0.243	-0.236	0.007	-0.247
7	-0.083	-0.047	0.036	-0.063	-0.057	0.006	-0.063
8	-0.334	-0.364	0.030	-0.382	-0.380	0.002	-0.365
9	-0.332	-0.375	0.043	-0.342	-0.338	0.004	-0.347
10	-0.217	-0.210	0.007	-0.215	-0.226	0.011	-0.217
TOTALS	-2.472	-2.470	0.294	-2.490	-2.474	0.060	-2.477
AVERAGES	-0.247	-0.247	0.029	-0.249	-0.247	0.006	0.303

	-0.247
SUM	-0.494
X-BAR A	-0.247
RANGE A	0.029

	-0.249
SUM	-0.496
X-BAR B	-0.248
RANGE B	0.006

RANGE VARIATION

RANGE A	0.029
RANGE B	0.006
SUM	0.035
R-BAR	0.018

X-BAR DIFFERENCE

X-BAR A	-0.247
X-BAR B	-0.248
X-BAR DIFF	0.001

PART RANGE (Rp)

Rp = 0.303

DEFINE TOTAL TOLERANCE: 1.00 mm

REPEATABILITY: EQUIPMENT VARIATION (EV)

EV= (R-BAR) x (K1)
EV= 0.081 mm

%EV / TOL.= 8.07%

%EV / TV= 16.24%

TRIALS	K1
2	4.56
3	3.05

REPODUCIBILITY: APPRAISER VARIATION (AV)

AV= SQRT{[(XDiff x K2)² - (EV² /nr)]}
AV= 0.018 mm

%AV / TOL.= 1.76%

%AV / TV= 3.54%

OPERATORS	K2
2	3.65
3	2.70

n = number of parts
r = number of trials
n x r = 20

REPEATABILITY & REPRODUCIBILITY (R & R)

R & R = SQRT{(EV)² + (AV)²}
R & R = 0.083 mm

% GAGE R & R / TOL.= 8.26%

%R & R / TV= 16.62%

PART VARIATION (PV)

PV= ABS(Rp x K3)
PV= 0.490 mm

%PV / TOL.= 49.01%

%PV / TV= 98.61%

PARTS	K3
2	3.65
3	2.70
4	2.30
5	2.08
6	1.93
7	1.82
8	1.74
9	1.67
10	1.62

TOTAL VARIATION (TV)

TV= SQRT{(GAGE R&R)² + (PV)² }
TV= 0.497 mm

%TV /TOL.= 49.70%

Figure 68 – Typical GR&R report per Measurement Systems Analysis Manual

ANOVA Gauge Repeatability & Reproducibility

R & R - ANOVA METHOD		Point ID	SPC 1
		Date	10/31/2007
		Page	

Part Number	Part Description	Part Type	Measurement Equipment Used	Location
9L34-5776-BA	SHKL ASY RR SPG			Art Eisenhauer

VALUES CONSIDERED IN THE R&R STUDY

Operators		PARTS										Averages	
		1	2	3	4	5	6	7	8	9	10		
Bob G.	1ª Reading	0.0355	0.0140	0.0095	0.0085	0.0120	0.0120	0.0140	0.0155	0.0085	0.0140	0.014	
	2ª Reading	0.0385	0.0140	0.0120	0.0105	0.0125	0.0115	0.0155	0.0155	0.0095	0.0130	0.015	
	3ª Reading	0.0390	0.0125	0.0105	0.0085	0.0130	0.0120	0.0140	0.0150	0.0085	0.0130	0.015	
	Average	0.038	0.014	0.011	0.009	0.013	0.012	0.015	0.015	0.009	0.013	Xa	0.015
	Range	0.004	0.002	0.003	0.002	0.001	0.001	0.002	0.001	0.001	0.001	Ra	0.002
Art E	1ª Reading	0.0380	0.0145	0.0120	0.0095	0.0135	0.0120	0.0140	0.0150	0.0090	0.0135	0.015	
	2ª Reading	0.0380	0.0130	0.0120	0.0080	0.0120	0.0125	0.0140	0.0155	0.0080	0.0120	0.015	
	3ª Reading	0.0380	0.0140	0.0095	0.0085	0.0130	0.0110	0.0145	0.0150	0.0085	0.0110	0.014	
	Average	0.038	0.014	0.011	0.009	0.013	0.012	0.014	0.015	0.009	0.012	Xb	0.015
	Range	0.000	0.002	0.003	0.002	0.002	0.002	0.001	0.001	0.001	0.003	Rb	0.001
Dan F.	1ª Reading	0.0375	0.0140	0.0100	0.0085	0.0135	0.0120	0.0130	0.0145	0.0085	0.0120	0.014	
	2ª Reading	0.0375	0.0145	0.0110	0.0100	0.0115	0.0115	0.0130	0.0150	0.0090	0.0120	0.015	
	3ª Reading	0.0295	0.0135	0.0095	0.0095	0.0120	0.0120	0.0140	0.0155	0.0080	0.0130	0.014	
	Average	0.035	0.014	0.010	0.009	0.012	0.012	0.013	0.015	0.009	0.012	Xc	0.014
	Range	0.008	0.001	0.002	0.002	0.002	0.001	0.001	0.001	0.001	0.001	Rc	0.002
Average of the Parts		0.037	0.014	0.011	0.009	0.013	0.012	0.014	0.015	0.009	0.013	Xbar	0.015
												Rp	0.028

Value of K1		Value of K2		Value of K3		D3 :	D4 :	A2 :		d₂:		XDif:	
0.59		0.5231		0.31		0.00	2.58	1.023		1.693		0.0010	

RANGE Average		Lower Control Limit		Upper Control Limit		Lower Control Limit		Upper Control Limit	
R_bar	0.002	LIC_Rbar =	0.000	LSC_Rbar =	0.0050	LIC_Xbar	0.012	LSC_Xbar	0.017

ANOVA CALCULATIONS

Data for Study		Origin of Variation	SS	DF	MS	F	CriticalFactor
# Parts	10		Sum of Squares	Degree of Freedom	Average Squares		
# Operators	3	OPERATOR	0.00001	2	0.000003	2.3955	
# Trials	3	PARTS	0.00534	9	0.000593	518.0054	*
Tolerance	0.10	APPRAISER by PART	0.00002	18	0.000001	0.8469	
		EQUIPMENT	0.00008	60	0.000001		
		TOTAL	0.00545	89			

Is the interaction of the operator part significant (F>CriticalFactor)?		No	Number of Distinct Categories	9

Estimate of Variance	Std. Dev.	% of Total Variation	% of Tolerance	% Contribution
Total Gage R&R	0.0012390	15.10%	7.43%	2.28%
Repeatability	0.0011424	13.9%	6.9%	1.9%
Reproducibility	0.0004797	5.8%	2.9%	0.3%
Operator (C4)	0.0002190	2.7%	1.3%	0.1%
Operator*Part (C4*C1)	0.0000000	0.0%	0.0%	0.0%
Part-To-Part	0.0081126	98.9%	48.7%	97.7%
Total Variation	0.0082066	100.0%	49.2%	

# of Data Categories	Tolerance Variation	Process Variation
APPROVED	APPROVED	APPROVED

GRAPHICAL ANALYSIS

APPROVAL

APPROVED BY	DATE

Figure 69 – ANOVA GR&R report

The example GR&R shown in figure 68, is the typical GR&R report for most OEM's. The ANOVA GR&R (figure 69) is also used by various OEM's, but should be used at different phases of the program.

Since most check fixtures or gages are bought off at the beginning of a program, the parts available for a GR&R have limited variation between them. While both GR&Rs (Mean and Range GR&R, and the ANOVA GR&R) provide a "percent of tolerance" error, the ANOVA method also provides a "percent of total variation". This variation is based on the parts, operators, measurement equipment, etc. in the study and because it provides a "percent of total variation" the resulting number can be deceiving.

In the ANOVA GR&R case, a GR&R result of greater than 30% in the "percent of total variation" doesn't necessarily mean the gauge is bad, what it indicates is that the gauge has a large percent of variation when compared to the overall variation.

For example, if the overall variation is 0.25mm and the GR&R percentage is 60%, that means (in a simple scenario) that the gauge with parts and operators has 0.15mm of error. While this may seem a lot if the gauge was checking a part with a 0.50mm tolerance, it is small if the tolerance is, say 1.50mm.

Additionally, since most of the parts used for an initial GR&R are prototype or first run parts, the chances of there being a large amount of variation between them are remote. This then can reduce the "number of distinct categories" to less than 5 which will result in a "failing" result. To overcome this, there are a few options, one being using a measurement device with increased decimal places to provide a possibility of more distinct categories another being getting some more parts for the study in hopes that they may provide larger variation.

With these things in mind, the author recommends using the ANOVA method:

1. In Production, where a build issue exists and an understanding of measurement variation versus process variation of the overall variation needs to be determined to focus on where an issue needs to be fixed.

2. Initial GR&R as a "percent of tolerance". This will eliminate the concerns of "little variation" part to part or not meeting the minimum distinct number of categories. It will also focus the viewer on understanding the overall measurement error versus the part tolerance. If this isn't done, there is the opportunity of "fixing" a good measurement system because the overall variation at the time of the study is small, thus potentially making the GR&R percentage a larger contributor.

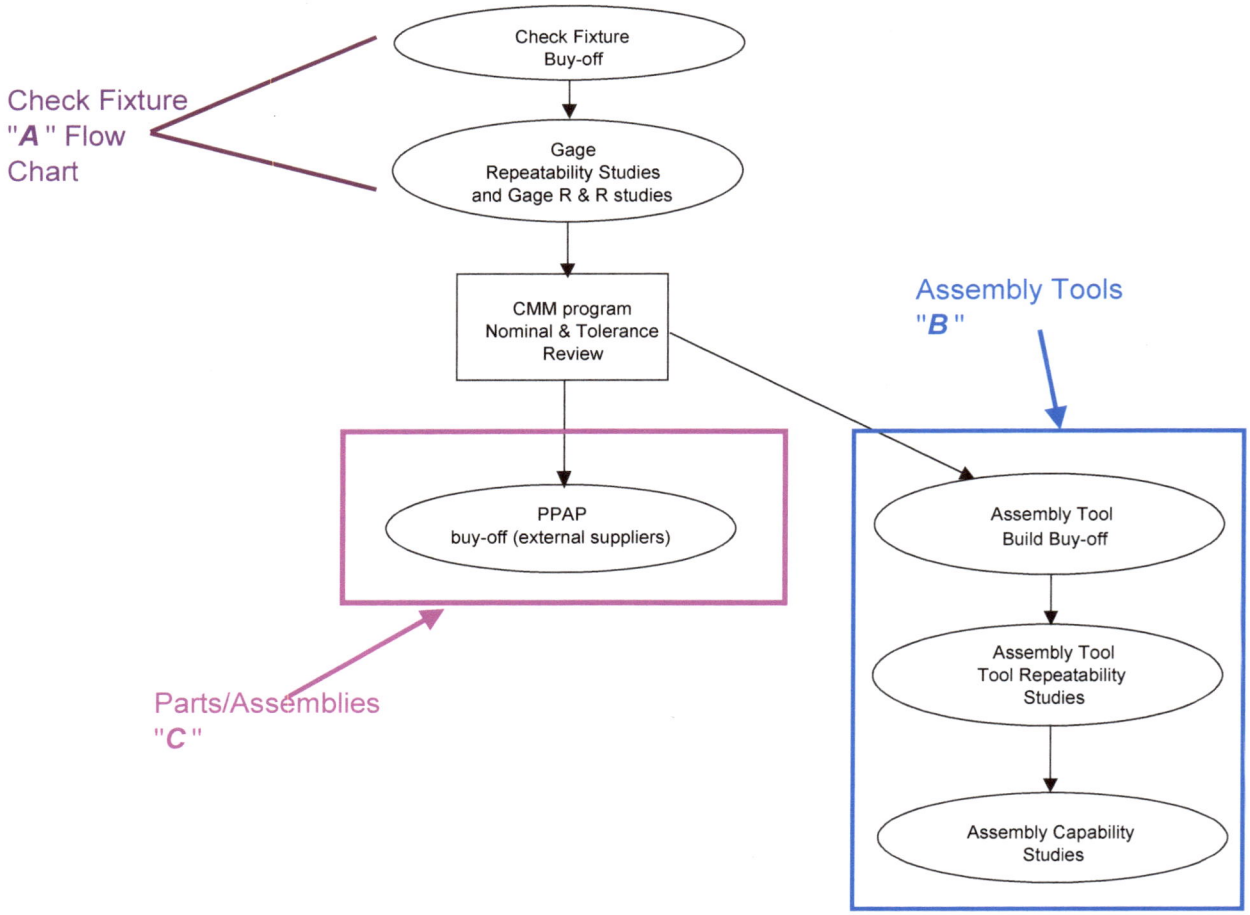

Check Fixture
"**A**" Flow
Chart

Check Fixture
Buy-off

Gage
Repeatability Studies
and Gage R & R studies

CMM program
Nominal & Tolerance
Review

Assembly Tools
"**B**"

PPAP
buy-off (external suppliers)

Parts/Assemblies
"**C**"

Assembly Tool
Build Buy-off

Assembly Tool
Tool Repeatability
Studies

Assembly Capability
Studies

CHECK FIXTURE FLOW CHART

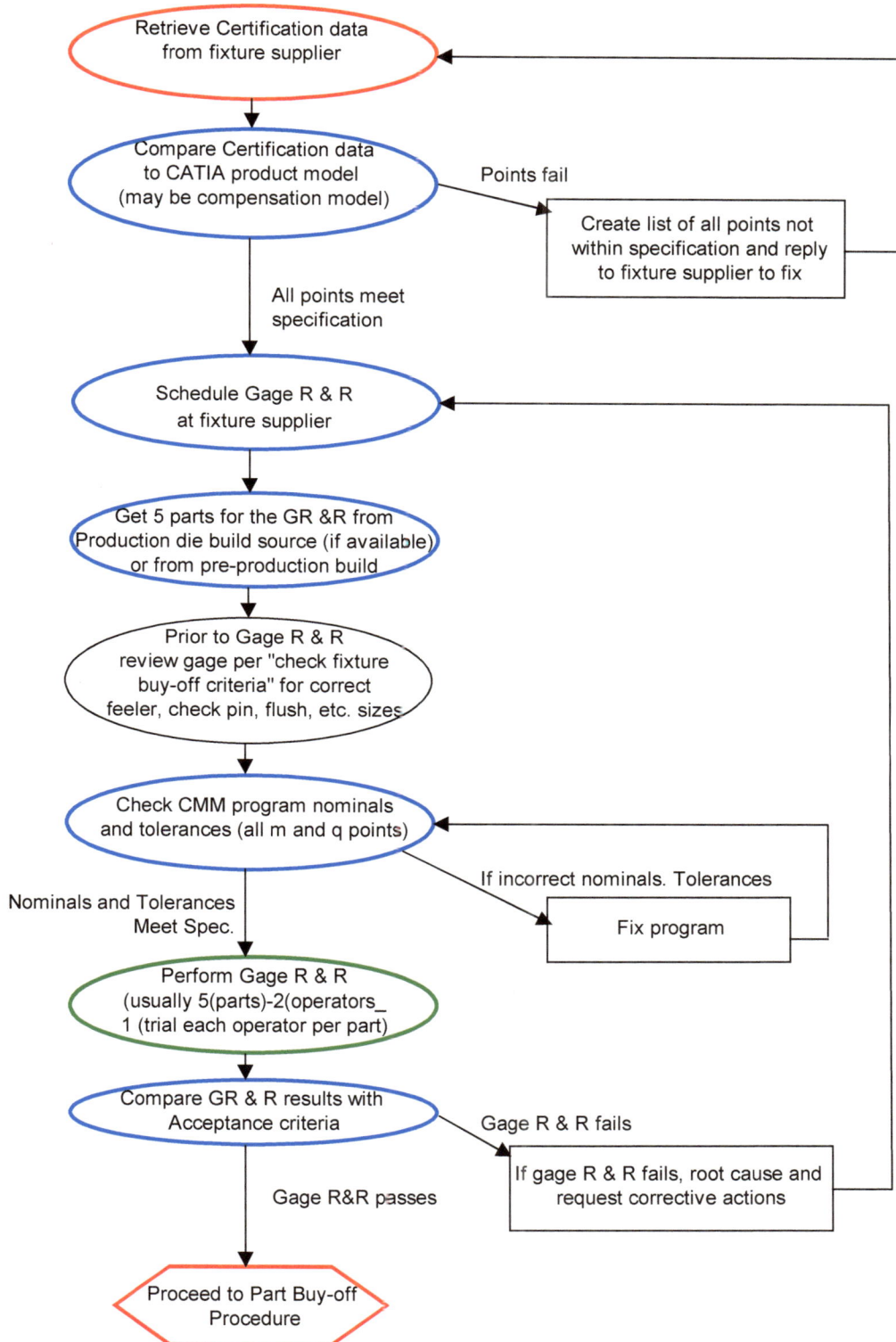

"**A**"

```
   ┌─────────────────────────┐
   │  Retrieve Certification  │ ◄───────────────────────────┐
   │  data from fixture       │                             │
   │  supplier                │                             │
   └───────────┬─────────────┘                             │
               │                                           │
               ▼                                           │
   ┌─────────────────────────┐                             │
   │  Compare Certification   │    Points fail              │
   │  data to CATIA product   │ ──────────►  ┌──────────────────────────┐
   │  model (may be           │              │ Create list of all points │
   │  compensation model)     │              │ not within specification  │─┘
   └───────────┬─────────────┘              │ and reply to fixture       │
               │                            │ supplier to fix            │
      All points meet                       └──────────────────────────┘
      specification
               │
               ▼
   ┌─────────────────────────┐
   │  Schedule Gage R & R     │ ◄───────────────────────────┐
   │  at fixture supplier     │                             │
   └───────────┬─────────────┘                             │
               │                                           │
               ▼                                           │
   ┌─────────────────────────┐                             │
   │  Get 5 parts for the GR  │                             │
   │  &R from Production die   │                             │
   │  build source (if        │                             │
   │  available) or from      │                             │
   │  pre-production build    │                             │
   └───────────┬─────────────┘                             │
               │                                           │
               ▼                                           │
   ┌─────────────────────────┐                             │
   │  Prior to Gage R & R     │                             │
   │  review gage per "check  │                             │
   │  fixture buy-off         │                             │
   │  criteria" for correct   │                             │
   │  feeler, check pin,      │                             │
   │  flush, etc. sizes       │                             │
   └───────────┬─────────────┘                             │
               │                                           │
               ▼                                           │
   ┌─────────────────────────┐                             │
   │  Check CMM program       │ ◄──────────┐                │
   │  nominals and tolerances │            │                │
   │  (all m and q points)    │            │                │
   └───────────┬─────────────┘            │                │
               │              If incorrect nominals.       │
      Nominals and Tolerances  Tolerances                  │
      Meet Spec.                           ┌──────────────┐│
               │                           │  Fix program │┘
               ▼                           └──────────────┘
   ┌─────────────────────────┐
   │  Perform Gage R & R      │
   │  (usually 5(parts)-2     │
   │  (operators_ 1 (trial    │
   │  each operator per part) │
   └───────────┬─────────────┘
               │
               ▼
   ┌─────────────────────────┐
   │  Compare GR & R results  │    Gage R & R fails
   │  with Acceptance         │ ──────────►  ┌──────────────────────────┐
   │  criteria                │              │ If gage R & R fails, root │─┘
   └───────────┬─────────────┘              │ cause and request         │
               │                            │ corrective actions        │
      Gage R&R passes                       └──────────────────────────┘
               │
               ▼
   ╱─────────────────────────╲
   │  Proceed to Part Buy-off  │
   │  Procedure                │
   ╲─────────────────────────╱
```

Production Gauge Build Review Checklist

1. Do all net pads match CAD (part) math model datum locations?

  - Do the corner net pads (where applicable) have clearance to the part in the corner?

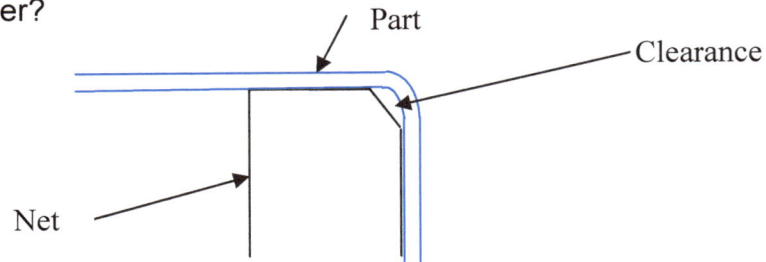

  - Do the corner clamps (where applicable) have clearance to the part in the corner?

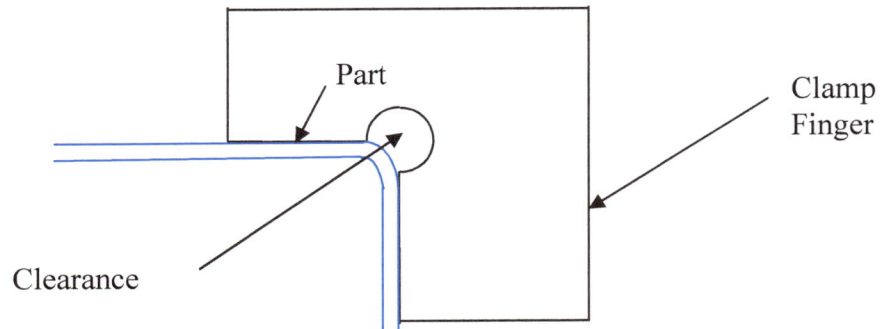

2. Do all pin locators match part math model datum locations?
3. Is gauge in assembly process position? (Floor line of gauge same as production process)
4. Is certification tag applied to gauge base and stamped?
5. Is GR&R tag applied to gauge base with appropriate information?
6. Are the gauge sequences of operation or Operator Description Sheets (ODS) attached to the fixture?
7. Do all clamps have a sequence number on them (does it match the sequence on the ODS sheets)?

Check fixture specific items

8. Do all critical measurement points have variable data capability?
    -   SPC bushing or JS block
9. Are the variable data points (SPC bushing/JS blocks) normal to surface?
10. Do all features of size have a check feature?
    -   Pin check for holes with 1.0mm or less positional tolerance
    -   Sight check for holes with greater than 1.0mm tolerance
    -   Pin check for size on locator holes
11. Is the check fixture to be used as a CMM holding fixture?
    -   Can the SPC/JS points be "dumped" out of the way of the CMM probe
    -   Can the flush/feeler rails be "dumped" out of the way of the CMM probe
    -   Does the part-to-base distance meet Check Fixture or Gauge Standards for CMM access underneath

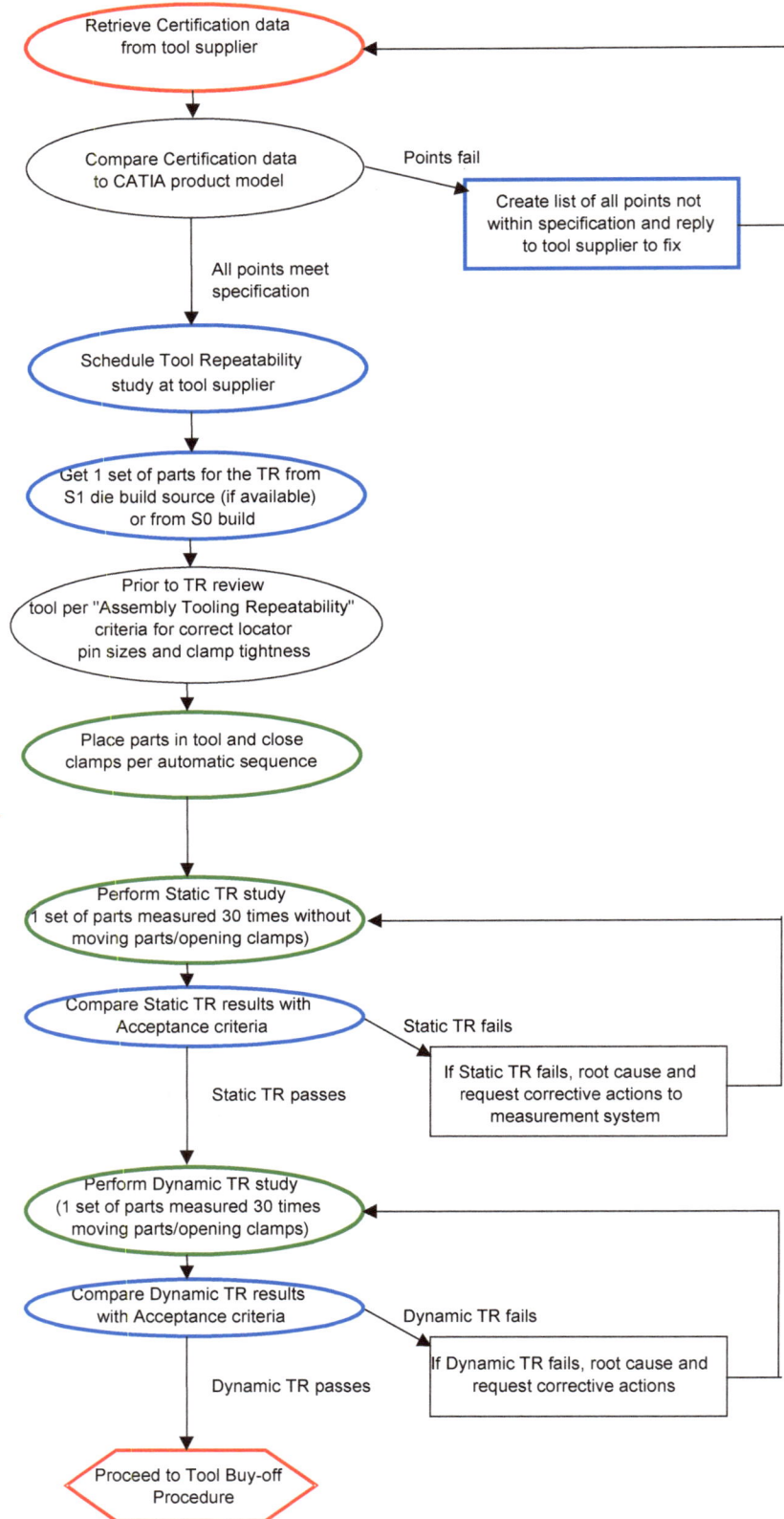

Chapter 6

"**B**"

Retrieve Certification data
from tool supplier

↓

Compare Certification data
to CATIA product model

→ Points fail →

Create list of all points not
within specification and reply
to tool supplier to fix

All points meet
specification

↓

Schedule Tool Repeatability
study at tool supplier

↓

Get 1 set of parts for the TR from
S1 die build source (if available)
or from S0 build

↓

Prior to TR review
tool per "Assembly Tooling Repeatability"
criteria for correct locator
pin sizes and clamp tightness

↓

Place parts in tool and close
clamps per automatic sequence

↓

Perform Static TR study
(1 set of parts measured 30 times without
moving parts/opening clamps)

↓

Compare Static TR results with
Acceptance criteria

→ Static TR fails →

If Static TR fails, root cause and
request corrective actions to
measurement system

Static TR passes

↓

Perform Dynamic TR study
(1 set of parts measured 30 times
moving parts/opening clamps)

↓

Compare Dynamic TR results
with Acceptance criteria

→ Dynamic TR fails →

If Dynamic TR fails, root cause and
request corrective actions

Dynamic TR passes

↓

Proceed to Tool Buy-off
Procedure

Below is an example of one assembly station within an assembly line. This particular station contains two operations

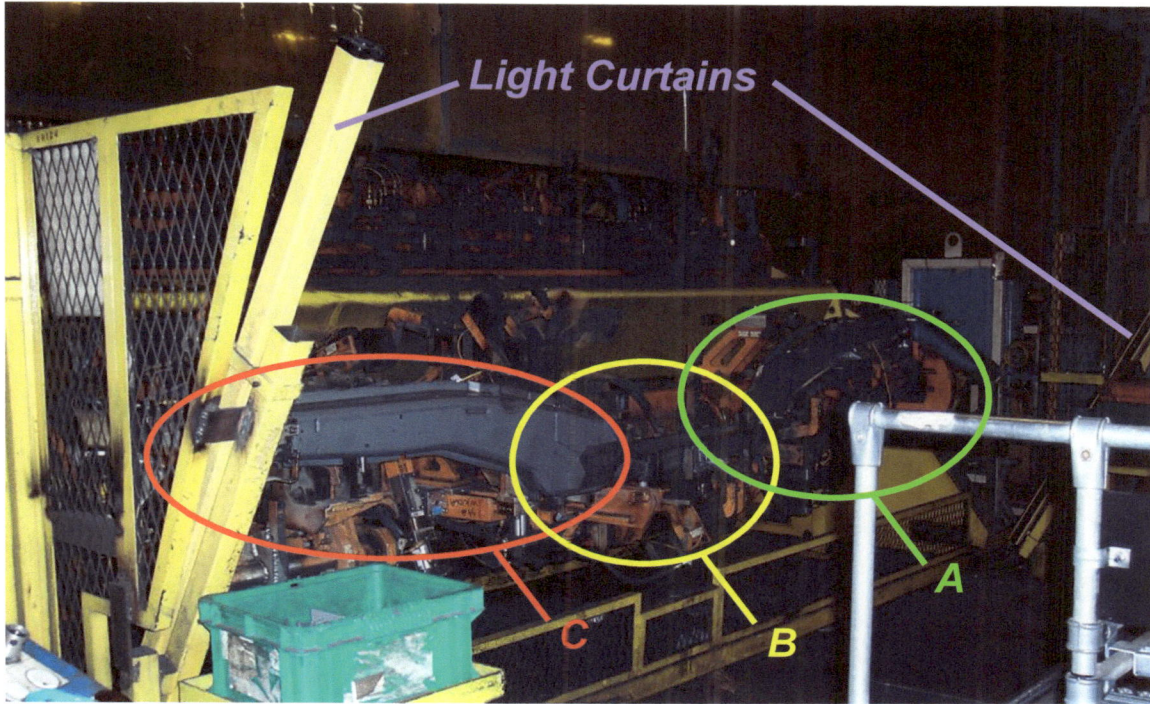

Figure 70 – Assembly Geometry setting station – 2 Operations

This station, like all others, has various safety requirements, from the fencing on the sides (yellow in the picture) and the light curtains (identified in figure 70) that shut down the station if the light is broken.

In Operation 10, below, there are two parts being welded together (part A and B).

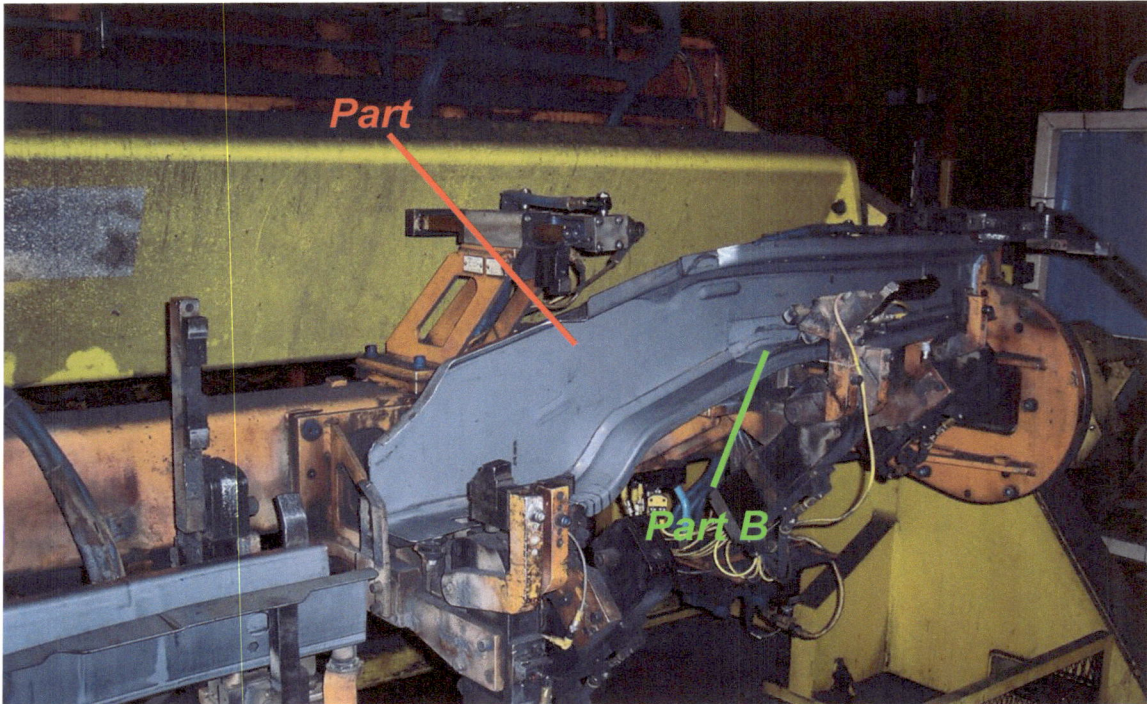

Figure 71 - Circle "A" – Operation 10

Here, part B is located to part A via form, or the two mating surfaces. A swing in clamp then holds the parts until the welds are completed. After this, the part then is hand loaded to operation 20 (see figure 70, circle C) and mated with another part (figure 72, circle B) for additional welding.

Below, figure 72, is a closer view of the assembly station showing greater detail and identifying where part locator pins, locator pad, clamps, and part presence sensors are.

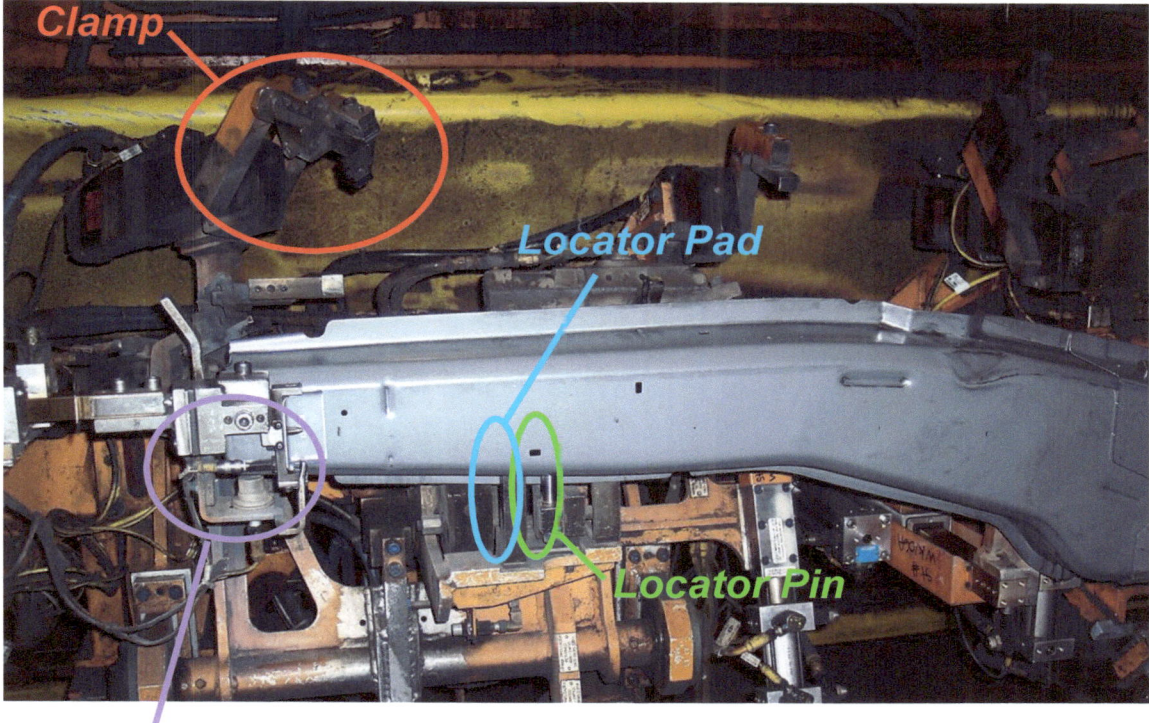

Figure 72 - Circle "C" – Operation 20

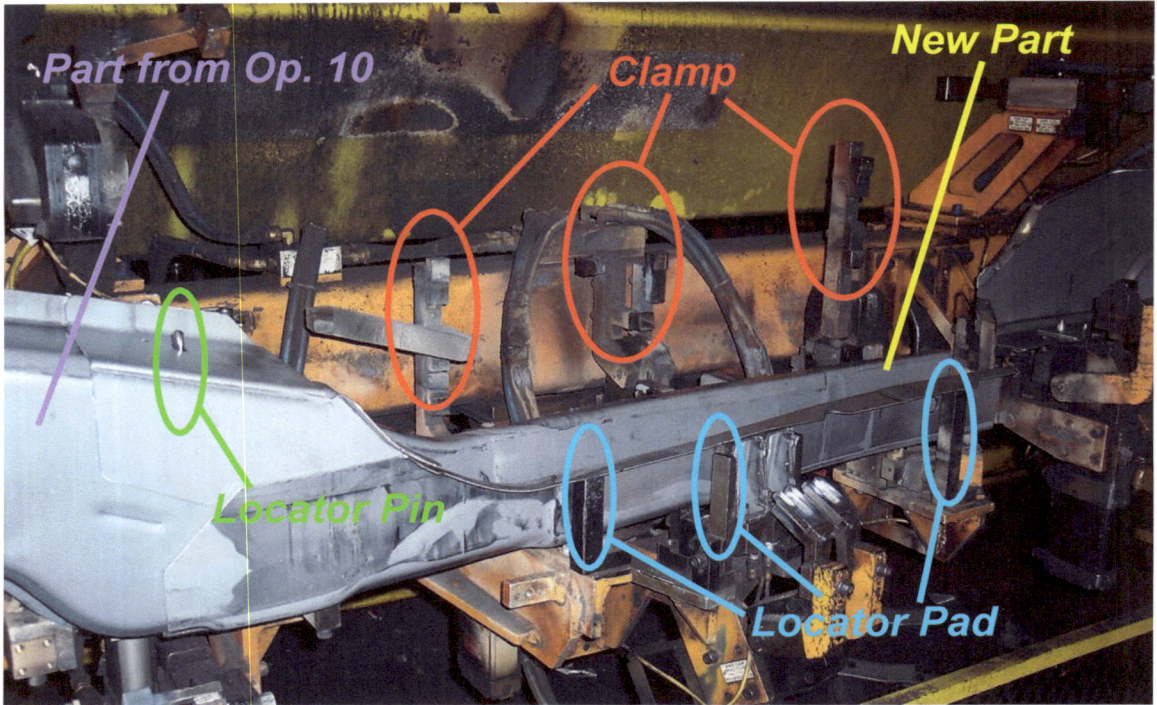

Figure 73 - Circle "B" – Operation 20

Assembly Tool Build Review Checklist

Without parts in the tool

1. Do all net pads match CAD math model locations?
    *   Do all corner net pads have clearance to the part in the corner?

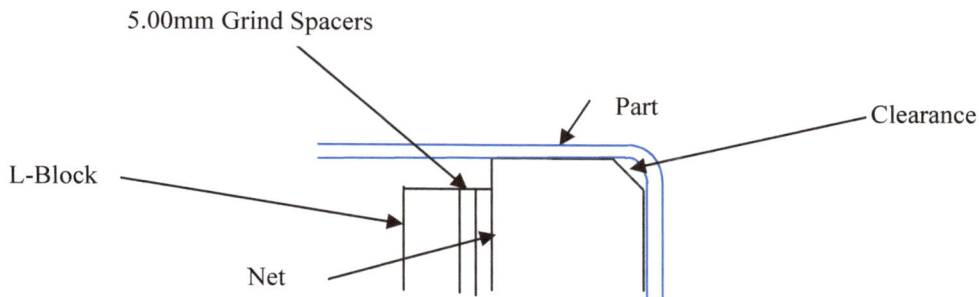

*   Do all corner clamp fingers have clearance to the part in the corner?

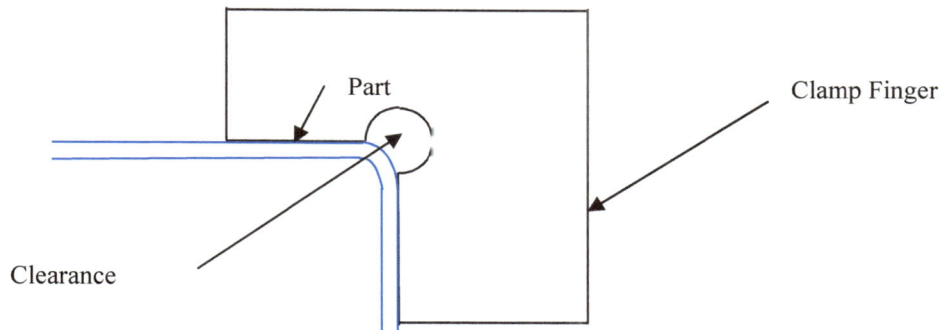

2. Do all pin locators match CAD math model locations?
3. Does the part match agreed to design orientation?
4. Is certification tag applied to gage base and stamped?
5. Are the assembly tool ODS (Operator Description Sheets) sheets attached to the fixture?
6. Do all clamps sequence in a similar order to the part gage?
7. Are all locator pin blocks doweled?
8. Are all locator pin mounting blocks shimmable in the directions of part locating?
9. Are all net blocks doweled?
10. Are all net blocks shimmable in the direction(s) of part locating?
11. Are the clamp fingers shimmable in the direction of clamp force?

With parts in the tool

1. Does any non-locating tooling contact the part (i.e. weld guns, proximity switches, etc.)?
2. Do the clamps have the proper clearance to the part (i.e. perform Feeler check)?
3. Are locating pins to part within specifications (i.e. perform caliper check)?
4. Do the tooling ball locations have body grid locations (i.e. X, Y, Z coordinates) stamped next to them?
5. Do the clamps have the ability to have the sequencing adjusted?

Shims – adjust pin
location in this direction

Shims – adjust pin
location in this
direction

Figure 74

Tooling pins, as shown above (figure 74), are shimmable in two directions. In most cases, the L-block, the pin mounting block, and the pin will be a NAAMS (North American Automotive Metric Standards) standard. The NAAMS standard pin is a shelf bought item, which enables the assembly/stamping plant to store and order quantities of them to improve maintenance abilities. At fixture build, two 5.00mm grind spacers are provided with the pin at nominal location. Additionally, the shim next to the L-block is tack welded to the L-block.

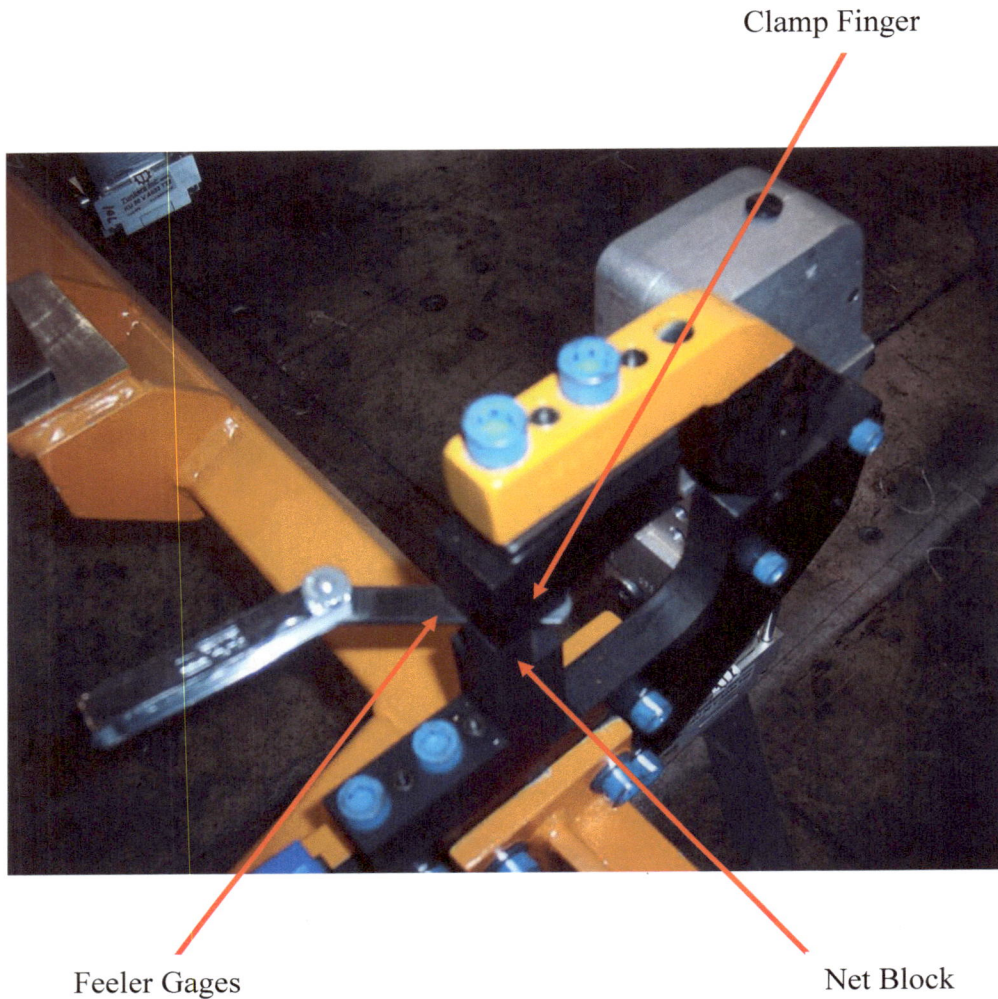

Clamp Finger

Feeler Gages

Net Block

Figure 75

Feeler Check – Clamp Finger to Net Block
Clamp finger pressure is verified by checking the distance between the clamp finger and the net block. This distance should be equivalent to the metal thickness(s) of the part(s) being held by the clamp depending on the clamps function.

Below is a standard guide for clamp to part or net block clearances that is used in the dimensional maintenance plan.

Clamp to net block clearance*
Net Clamp   up to 0.12mm clear
Slip Clamp   0.12mm to 0.25mm clear

*Denotes that these values are in addition to the metal thickness of the part itself.

A net clamp condition is that of a clamp that holds the part to the net pad without allowing the part to move.

The slip clamp provides clearance for the part to move where clamps may be bringing the part to the net pad in another direction. An example of this is when clamps are sequenced to close in the fore/aft direction; but the part also must be brought to the nets in the up/down direction.

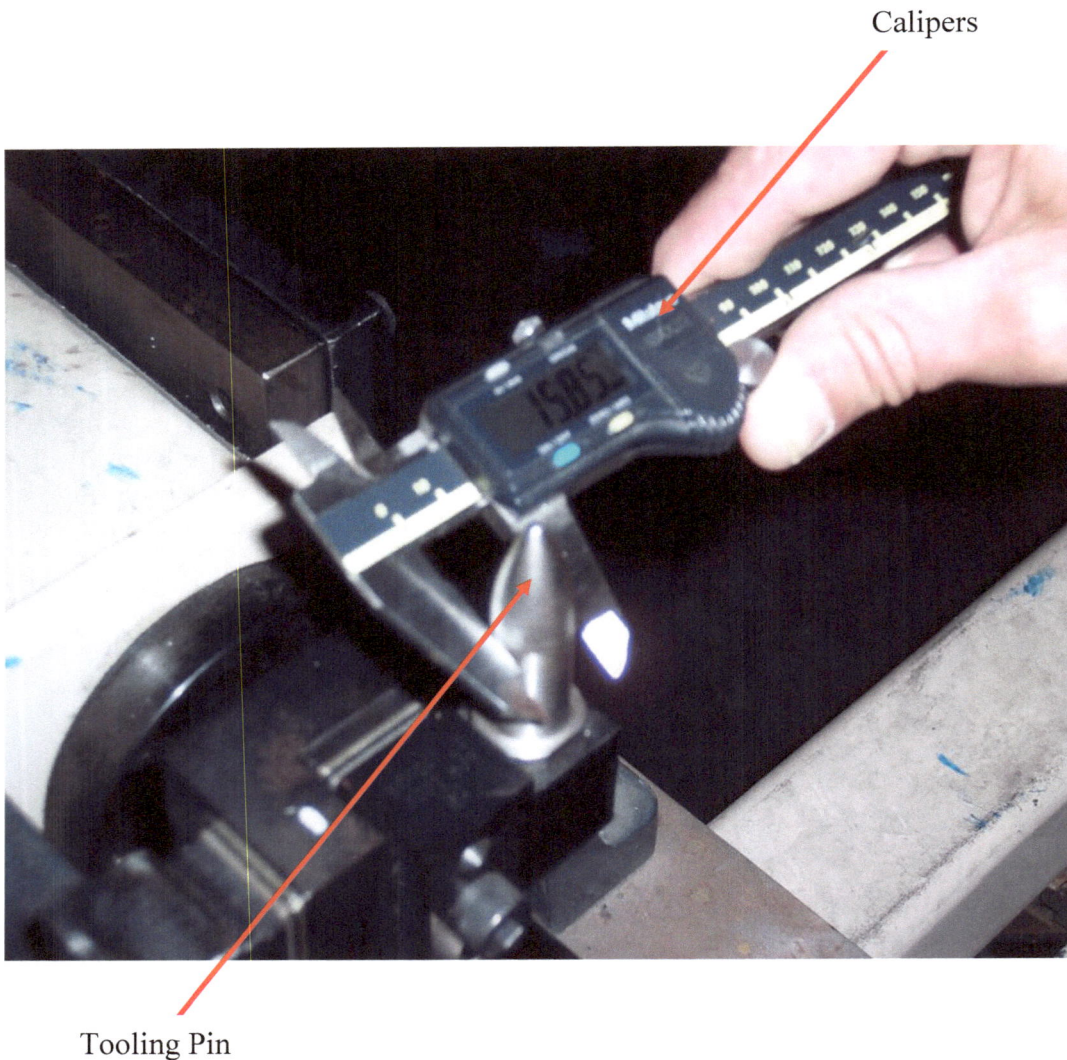

Calipers

Tooling Pin

Figure 76

Tooling Pin – Size check
Tooling locator pins also require a size check in addition to the certification of body grid location. Calipers are used to perform this check for both the pin and the respective hole/slot in the mating part.

A standard when comparing part to pin clearance is the following:

Part Hole/Slot LMC

$$\left[\begin{array}{l} \text{Nominal Part Hole/Slot size} \\ + \quad \text{Maximum hole/slot size tolerance} \end{array}\right]$$

-   NAAMS standard for minimum pin sizes

$$\left[\begin{array}{l} \text{Nominal Part Hole/Slot size} \\ - \quad 0.15 \text{ mm NAAMS nominal pin size} \\ - \quad 0.02\text{mm size tolerance} \end{array}\right]$$

Prior to Production Launch

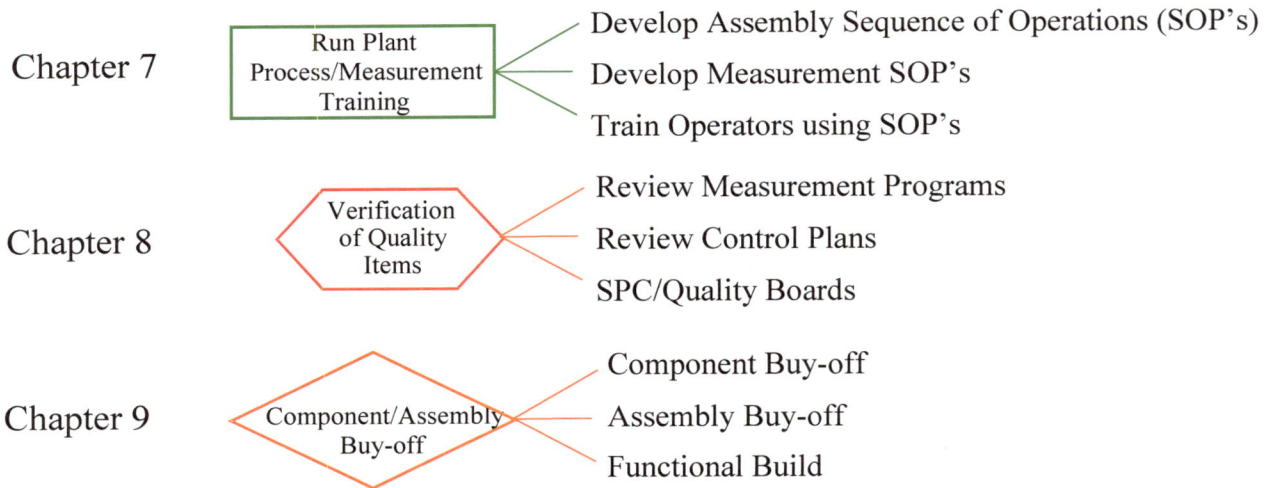

Chapter 7 — Run Plant Process/Measurement Training
- Develop Assembly Sequence of Operations (SOP's)
- Develop Measurement SOP's
- Train Operators using SOP's

Chapter 8 — Verification of Quality Items
- Review Measurement Programs
- Review Control Plans
- SPC/Quality Boards

Chapter 9 — Component/Assembly Buy-off
- Component Buy-off
- Assembly Buy-off
- Functional Build

Everything discussed thus far with regards to repeatability and certification, use of a gauge, assembly tool, die, etc. assumes that a specific set of instructions is followed. It is also imperative that when performing various types of repeatability studies, a sequence of operations (SOP) be documented and followed for each trial and every operator. This provides a one avenue of change if the repeatability study fails; or, provides a common approach for future measurement if the repeatability study passes.

Example of process sequence of operations

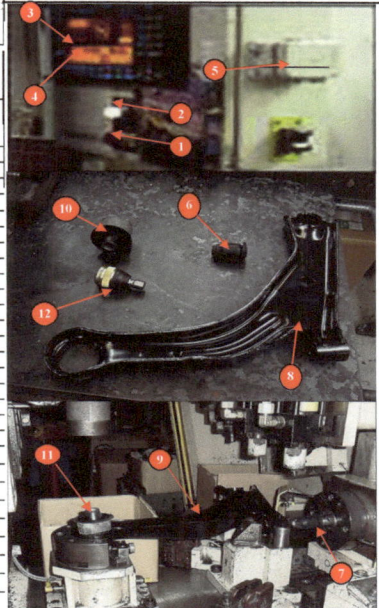

QUALITY PROCESS SHEET	DEPARTMENT:	1012		Q.P.S Sheet Number:	of		Date	Authorized By:
	SECTION:	N/A						

PLANT NAME: | PROCESS:

OPERATION
J61121 Final Assembly

Part Name: J61121
Base Part Number (if required): SEE BELOW

REFERENCE CBOM for correct part number. Reference Production Process Controls & Reaction Plans.

TARGET TIME:		CYCLE TIME:							
K	IN PROCESS STOCK	SAFETY & ERGO	DELTA CRITICAL	QUALITY CHECK	QUICK CHANGE OVER	VISUAL FACTORY	ERROR PROOFING		
E									
Y									

STEP	WORK STEPS – ALWAYS FOLLOW SAFE WORK PRACTICES	KEY	POSSIBLE HAZARDS	RECOMMENDED SAFE JOB PROCEDURE
	Initial Power-up of the System			
1)	Pull Emergency Stop Button			
2)	Push Control power on button			
3)	Push hydraulic pump key on.			
4)	Panel view (F2) - push reset			
5)	Panel view (F3) - push suspend key home			
	*All cylinders key (F4) should have a green and blue light on the status lights above the panel view. Look at panel view to see what canister is selected to run LH part - 70601 (Tower part#) - 2L84 3A 424 - Ford Part Number)			
	Part Operation Instructions			
6)	Inspect rubber bushing for voids, missing o-ring, and disconfiguration of flange.	◇		
7)	Insert rubber bushing on axis #4 guide pin with flange towards the axis.			
8)	Inspect control arm for splits, bad paint, gouges or any disconfiguration.	◇		
9)	Insert control arm in machine with flange facing up and tube against stop.			
10)	Inspect canister bushing for bad rubber, rust, sizing marks, and any disconfiguration.	◇		
11)	Insert canister bushing on canister guide pin with flange up and rubber slot aligned with the slot guide.			
12)	Inspect ball joint for torn rubber boot, and rubber boot not seated to flange.	◇		

PERSONAL PROTECTIVE EQUIPMENT	SAFETY LOCKS	SAFETY GLASSES	SAFETY SHOES	GLOVES	HEARING PROTECTION	HARD HAT	FACE MASK	FACE MASK	APRON	SLEEVES
See other side for PPE	NO	YES	YES	YES	YES	NO	NO	Half Face Resp. NO	OPTIONAL	NO

Sign-off / Reps.	1st	2nd	3rd
Operators			
Workgroup Leader			
MOS			
Health & Safety			N/A

Figure 77 – Process Sequence of Operations

Here, not only is a detailed sequence outlined so anyone can perform the operation, but specific symbols highlight key steps where inspection occurs and what type of safety equipment is needed for this process. Additionally, pictures are provided with numbered balloons that coincide with the procedural descriptions defined on the left.

Developing Measurement SOPs

Measurement SOPs are a combination of two elements: the control plan and the sequence of operation of the gauge or check fixture.

Similar to the process sequence of operations, the gauge SOPs detail specifically how to use the gauge. From identifying the clamping sequence and locator pins used to achieve a passing gage R&R, to what details are used, where to perform pin checks, SPC checks, etc.

Additionally, the SOPs will document the following: type of check, from a visual inspection of a hole, to an indicator check of a specific point, to a template check of a surface or edge. These check types should match what is identified on the control plan for the part or assembly being checked.

When developing a SOP for a measurement device, you should follow the process of how you will check the part without overloading the SOP. The theory is that too much information will make it difficult to read or follow.

Typically, the following process has been followed by some NA OEM's and Tier One suppliers:

> Sheet 1 – Locator and Clamp Sequences identified
> Sheet 2 – Feeler check and trim check locations
> Sheet 3 – Hole and slot check locations
> Sheet 4 – Variable indicator check locations

Below is an example of these types of SOPs.

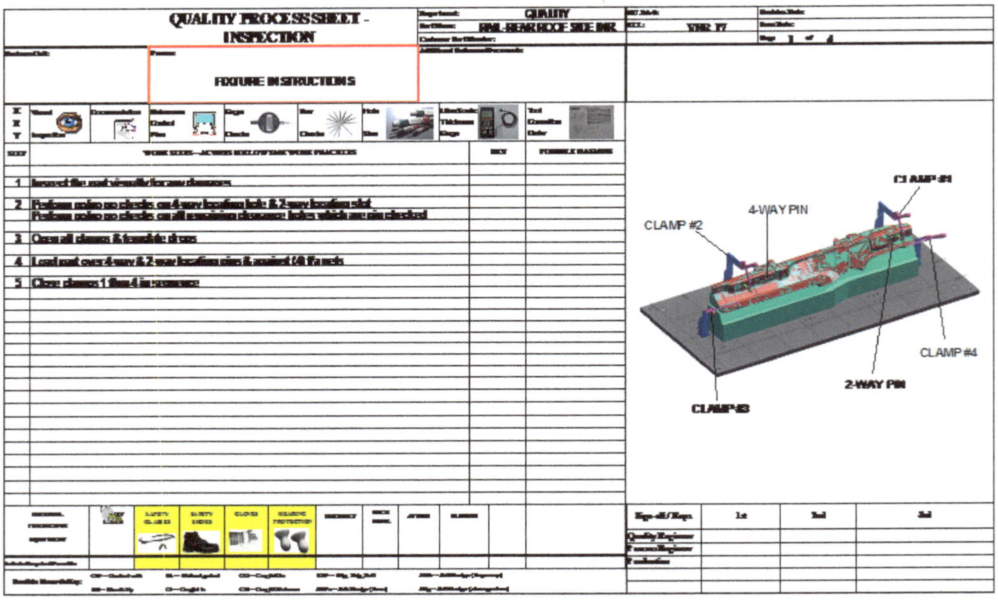

Figure 78

Figure 78 is an example of sheet 1 detailing the clamping sequence and location of all datums, clamps, and locator pins.

After the sequences have been identified, training of the specific operators is needed to ensure that the operator understands the requirements of the operation, can perform the requirements, understands the importance of not only following the sequences consistently, and identifies to their superior. Any non-conformance they encounter can be addressed immediately.

Training does not end upon completion of the initial training phase. It is recommended that each operator be re-certified based on a time interval, or prior to a job change, to ensure that the operator understands and is able to perform the functions required of him/her. These training sessions should be documented as per the requirements of TS-16949 or other internal requirements.

Finally, if the operator(s) using the check fixture or gauge are different than the ones that performed the initial GR&R, the GR&R should be redone to understand the amount of influence this specific operator will have on the overall measurement, and if needed provide further training and documentation to improve the measurement system results.

Chapter 8 - Verification of quality related items.

In this section we will identify areas that need to be verified and the steps needed to ensure the accuracy of using these items.

Items needing verification:
1. Measurement Programs
2. Measurement fixtures (gages)
3. Assembly tooling

1. Measurement Programs
   While there are multiple systems to program measurement, such as: mig weld systems, datamytes, etc., this section will focus on CMM measurement programs. By focusing on and identifying various items in a CMM program, we can use this information and procedure when reviewing programs from other measurement methods.

   Measurement programs need to be verified prior to any measurement to ensure accuracy in the results reported. Verification must be done on the minimum items:
   a. Part Number
   b. Part Engineering Change level
   c. Point ID's
   d. Point Nominals
   e. Tolerances
   f. Vectors
   g. Material Thickness and direction
   h. Alignment method
   i. Data outputs – directions of the measurement results that will be shown.

Figure 79 is an example of a PCDMIS CMM program printout. This printout shows the part information, how the CMM will align to get into the part coordinate system, how it will move from one measurement location to the next, what measurement features to hit and what to output to a data file.

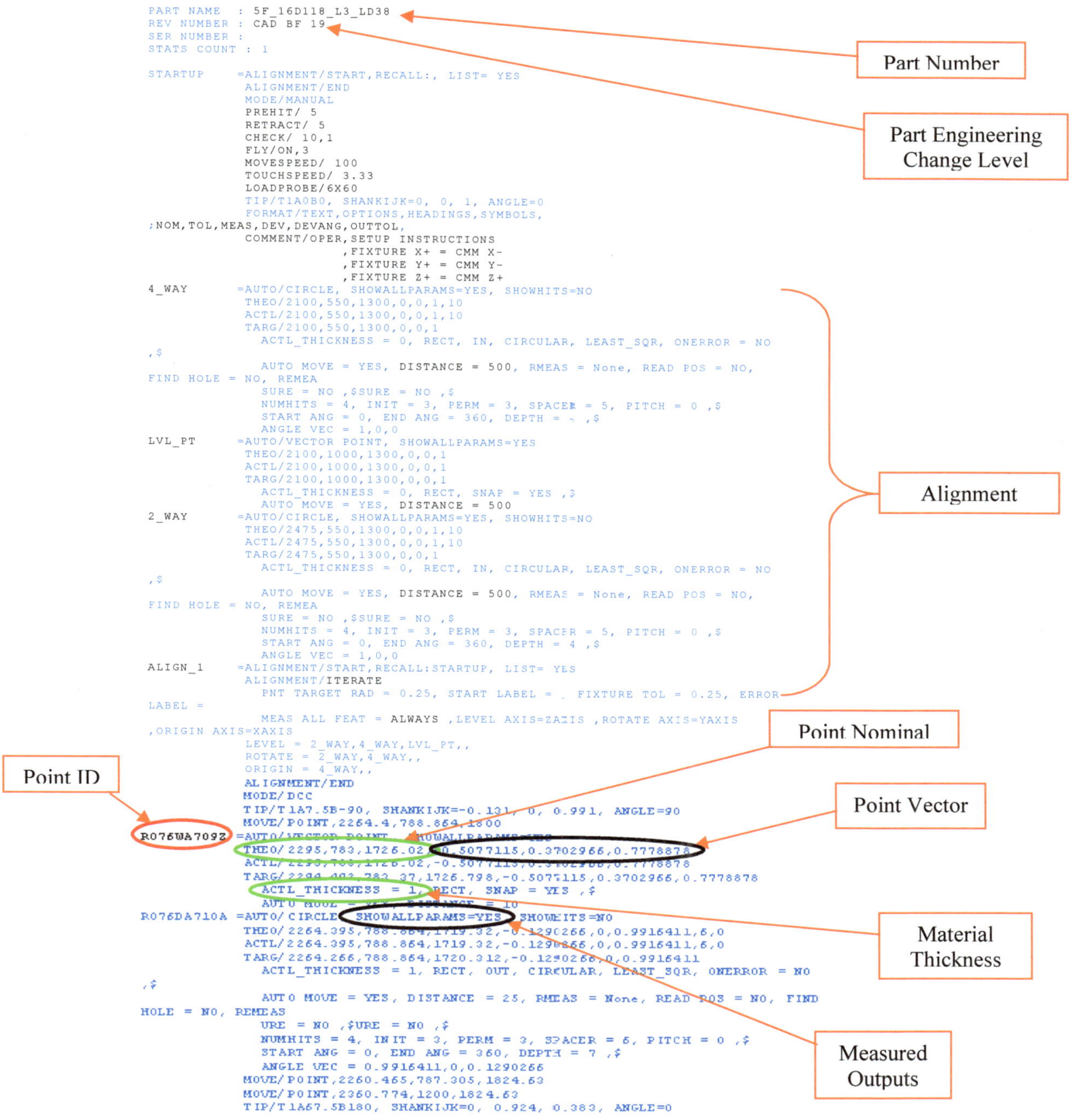

```
PART NAME   : 5F_16D118_L3_LD38
REV NUMBER  : CAD BF 19
SER NUMBER  :
STATS COUNT : 1

STARTUP    =ALIGNMENT/START,RECALL:, LIST= YES
           ALIGNMENT/END
           MODE/MANUAL
           PREHIT/ 5
           RETRACT/ 5
           CHECK/ 10,1
           FLY/ON,3
           MOVESPEED/ 100
           TOUCHSPEED/ 3.33
           LOADPROBE/6X60
           TIP/T1A0B0, SHANKIJK=0, 0, 1, ANGLE=0
           FORMAT/TEXT,OPTIONS,HEADINGS,SYMBOLS,
;NOM,TOL,MEAS,DEV,DEVANG,OUTTOL,
           COMMENT/OPER,SETUP INSTRUCTIONS
                        ,FIXTURE X+ = CMM X-
                        ,FIXTURE Y+ = CMM Y-
                        ,FIXTURE Z+ = CMM Z+
4_WAY      =AUTO/CIRCLE, SHOWALLPARAMS=YES, SHOWHITS=NO
           THEO/2100,550,1300,0,0,1,10
           ACTL/2100,550,1300,0,0,1,10
           TARG/2100,550,1300,0,0,1
                ACTL_THICKNESS = 0, RECT, IN, CIRCULAR, LEAST_SQR, ONERROR = NO
,$
                AUTO MOVE = YES, DISTANCE = 500, RMEAS = None, READ POS = NO,
FIND HOLE = NO, REMEA
                SURE = NO ,$SURE = NO ,$
                NUMHITS = 4, INIT = 3, PERM = 3, SPACER = 5, PITCH = 0 ,$
                START ANG = 0, END ANG = 360, DEPTH = 4 ,$
                ANGLE VEC = 1,0,0
LVL_PT     =AUTO/VECTOR POINT, SHOWALLPARAMS=YES
           THEO/2100,1000,1300,0,0,1
           ACTL/2100,1000,1300,0,0,1
           TARG/2100,1000,1300,0,0,1
                ACTL_THICKNESS = 0, RECT, SNAP = YES ,$
                AUTO MOVE = YES, DISTANCE = 500
2_WAY      =AUTO/CIRCLE, SHOWALLPARAMS=YES, SHOWHITS=NO
           THEO/2475,550,1300,0,0,1,10
           ACTL/2475,550,1300,0,0,1,10
           TARG/2475,550,1300,0,0,1
                ACTL_THICKNESS = 0, RECT, IN, CIRCULAR, LEAST_SQR, ONERROR = NO
,$
                AUTO MOVE = YES, DISTANCE = 500, RMEAS = None, READ POS = NO,
FIND HOLE = NO, REMEA
                SURE = NO ,$SURE = NO ,$
                NUMHITS = 4, INIT = 3, PERM = 3, SPACER = 5, PITCH = 0 ,$
                START ANG = 0, END ANG = 360, DEPTH = 4 ,$
                ANGLE VEC = 1,0,0
ALIGN_1    =ALIGNMENT/START,RECALL:STARTUP, LIST= YES
           ALIGNMENT/ITERATE
                PNT TARGET RAD = 0.25, START LABEL = , FIXTURE TOL = 0.25, ERROR
LABEL =
                MEAS ALL FEAT = ALWAYS ,LEVEL AXIS=ZAXIS ,ROTATE AXIS=YAXIS
,ORIGIN AXIS=XAXIS
                LEVEL = 2_WAY,4_WAY,LVL_PT,,
                ROTATE = 2_WAY,4_WAY,,
                ORIGIN = 4_WAY,,
           ALIGNMENT/END
           MODE/DCC
           TIP/T1A7.5B-90, SHANKIJK=-0.131, 0, 0.991, ANGLE=90
           MOVE/POINT,2264.4,788.864,1800
R076WA7092 =AUTO/VECTOR POINT, SHOWALLPARAMS=YES
           THEO/2295,783,1726.02,-0.5077115,0.3702966,0.7778878
           ACTL/2295,783,1726.02,-0.5077115,0.3702966,0.7778878
           TARG/2294.003,782.37,1726.798,-0.5077115,0.3702966,0.7778878
                ACTL_THICKNESS = 1, RECT, SNAP = YES ,$
                AUTO MOVE = YES, DISTANCE = 10
R076DA710A =AUTO/CIRCLE, SHOWALLPARAMS=YES, SHOWHITS=NO
           THEO/2264.395,788.864,1719.32,-0.1290266,0,0.9916411,6,0
           ACTL/2264.395,788.864,1719.32,-0.1290266,0,0.9916411,6,0
           TARG/2264.266,788.864,1720.312,-0.1290266,0,0.9916411
                ACTL_THICKNESS = 1, RECT, OUT, CIRCULAR, LEAST_SQR, ONERROR = NO
,$
                AUTO MOVE = YES, DISTANCE = 25, RMEAS = None, READ POS = NO, FIND
HOLE = NO, REMEAS
                URE = NO ,$URE = NO ,$
                NUMHITS = 4, INIT = 3, PERM = 3, SPACER = 6, PITCH = 0 ,$
                START ANG = 0, END ANG = 360, DEPTH = 7 ,$
                ANGLE VEC = 0.9916411,0,0.1290266
           MOVE/POINT,2260.465,787.305,1824.63
           MOVE/POINT,2360.774,1200,1824.63
           TIP/T1A67.5B180, SHANKIJK=0, 0.924, 0.383, ANGLE=0
```

Part Number

Part Engineering Change Level

Alignment

Point Nominal

Point Vector

Point ID

Material Thickness

Measured Outputs

Figure 79

- 114 -

While the part number, part engineering change level, and point IDs are obvious in understanding point nominals; vectors may be slightly harder. In the simplest terms, the point nominal is the location in 3-Dimensional space that the point exists; usually an X, Y and Z coordinate. The Vector of the point is the 3D normal line created from the surface that the point lies on and through the point – see figure 80. The vector is important, as identified in chapter 3, figure 32, so that measurement error can be reduced where possible.

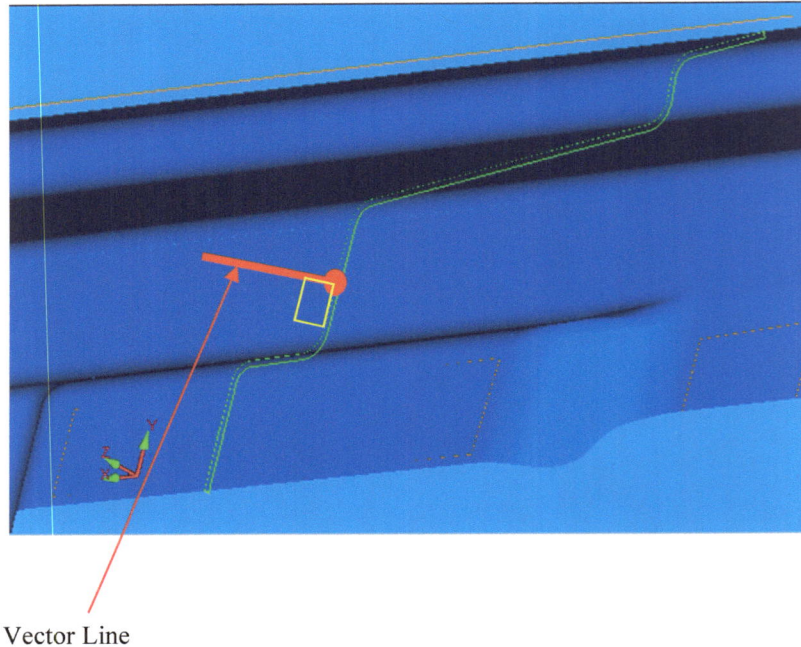

Vector Line

Figure 80

Material thickness and direction is also important when measuring a point, these items might effect the deviation produced. In most automotive CAD designs, the part is created with no thickness in the 3D model itself, commonly referred to as "design side". Various CAD systems allow the CAD designer/engineer to apply a feature to the surface identifying it with a thickness and direction. This can be accomplished in a few ways:

Direction –
   a. In a 2D section that resides in the 3D model showing the part thickness (see figure 81).
   b. An arrow in the 3D model attached to the specific surface indicating which direction material goes.

Thickness –

c. In a 2D section that resides on the 3D model showing the part thickness (see figure 81).
d. As a reported feature in the identification of the part/surface selected (see figure 82).
e. In the "title block" of the 2D drawing of the part.

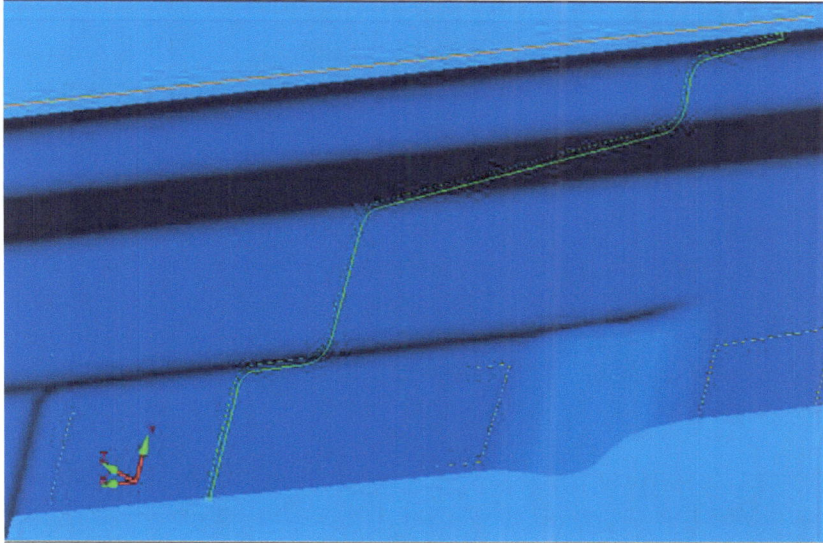

Figure 81

Figure 82

Alignment method will identify which features are used to get the measurement device, in the same coordinate system, as the item to be measured. Some alignment methods are:

1. *3-2-1 alignments* – this alignment uses the part datums (3 points for a plane, 2 for a line, 1 for a point) to set up the coordinate system. The measurement device, in this example a CMM, will measure these points and then use the input reference values to establish the coordinate system.
2. *Fixture alignment* – similar to the 3-2-1 alignments as it uses the same concept; but instead of using the part to set up the coordinate system, it uses the check fixture or gauge benchmarks (locations usually on the base that were used to certify and machine the gauge). See figure 76 for an example.

While these methods are the most widely used in the automotive arena, there are others. In addition, the alignment on a CMM for features of size is typically based on RFS (Regardless of Feature Size) and not MMC or LMC. This may cause differences in capability and repeatability results on items identified in the GD&T feature control frames as MMC or LMC.

Data outputs are directions of the measurement results that will be shown. When a CMM measures a point, all directions can be output as data, though some of these directions may have a deviation of "zero" because it is a set location of where the CMM hits the point (see figure 83 for examples).

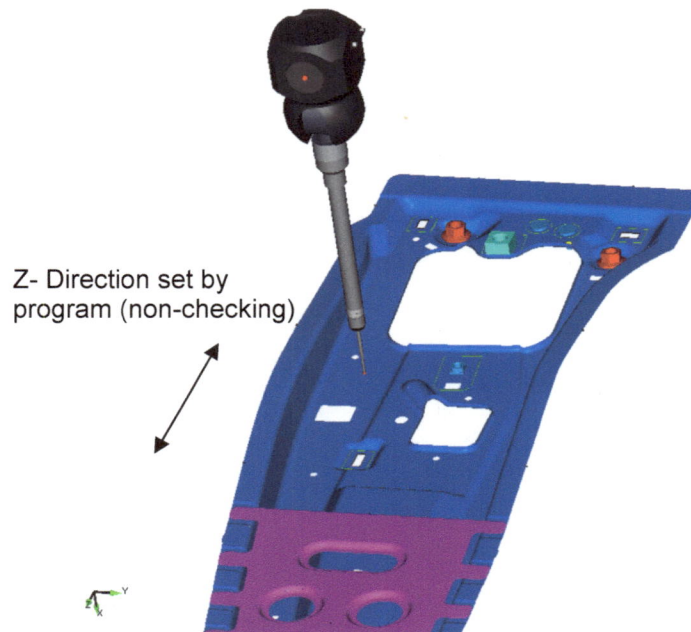

Z- Direction set by program (non-checking)

Figure 83 – Surface measurement

In figure 83, the CMM is directed to a specific point to measure in the "Z" direction, so the data output for the "X" direction would have a zero deviation as that is the move or set direction. The "Y" direction may also have a zero deviation, depending on whether the surface measured is parallel to the Y plane or not.

Features of size may be output by each individual direction specified or a "positional" location, which combines all directions and reports a vector deviation back to the nominal (see figure 84) for graphic of this method). Note: that the positional location will depend on where the nominal point resides.

DIM D14 = LOCATION OF SURFACE POINT S2659M3001A

AX	NOMINAL	MEAS	+TOL	-TOL	DEV	OUTTOL
X	1307.887	1307.887	0.050	0.050	0.000	0.000
Y	-470.345	-470.321	0.050	0.050	0.024	0.000
Z	1272.725	1272.728	0.050	0.050	0.003	0.000
T	0.000	0.024	0.050	0.050	0.024	0.000

DIM D23 = LOCATION OF CIRCLE H2659M3009A

AX	NOMINAL	MEAS	+TOL	-TOL	DEV	OUTTOL
X	1500.000	1498.231	0.500	0.500	1.769	1.269
Y	-475.000	-474.654	0.500	0.500	-0.346	0.000
Z	1272.725	1272.728	0.050	0.050	0.003	0.000
D	12.000	12.153	0.250	0.250	0.153	0.000
T	0.000	1.803	0.500	0.500	1.803	1.303

Figure 84

While this is not an issue in most cases, a measurement of an extrusion may be cause of concern. Since the measured point is based on creating a "plane" for one direction, that plane will usually reside on the surface that the extrusion starts from. The feature of size is measured, and then the center of the feature is normally projected to the plane. In the case of an extrusion, the point in the CAD model may not reside on the surrounding surface but at the end of the extrusion. If this is the case, then there may be a triangulation error in the CMM projected measurement, a deviation error as the point measured is on one surface versus the nominal on another (see Point B, figure 82). This may be overcome by assigning an offset to the measured plane, based on the theoretical distance from the plane to the extrusion point specified, though this could also have some measurement error.

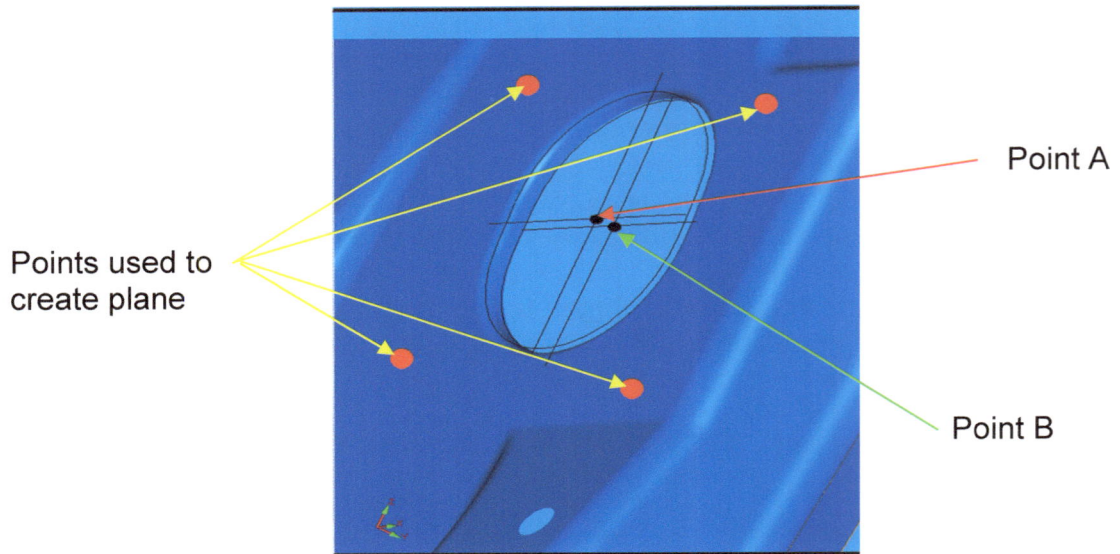

Figure 85

Now that we have the measurement information covered, let us look at the check fixture or gauge that the part will be verified on. The objective is to ensure that the part is in specification.

Over the next few pages, enclosed are checklists of items to review on a check fixture or gauge. These items, if verified, will ensure that the fixture meets all certification and physical requirements. So, that information gathered when checking a part is accurate and repeatable.

Figure 86 – Check Fixture

The check fixture or gauge in figure 86 is a full periphery check with four clamps (defined by the red handles with yellow washers), 2 locator pins, including various feeler, pin, and template checks. The yellow washers are used as a safety feature so the operator's hand will not get "pinched". This fixture is, what would typically be described as a "full content" fixture. It contains "drop templates (see item A in figure 86, previous page) to check the surfaces of the downstanding faces and supply a bushing for an SPC measurement location, feeler pins – made to "go/no go" tolerance specifications (see item B in figure 86, previous page), and check pins for hole size and position checks (see item C in figure 86, previous page).

Another type of fixture, known as a "De-contented" fixture (see figures 87 and 95), is a fixture with no added content. While there is typically a feeler and flush surface, there are no feeler blocks to check it with. Hole checks are performed via sight, and SPC checks are completed using a hand held data collector the operator positions themselves at a rough location identified on the fixture. For this fixture, an operator may use a "taper gage" (see figure 89) for surface checks and gage pins to check hole sizes. All of these items are typically standard equipment in a quality lab.

Figure 87

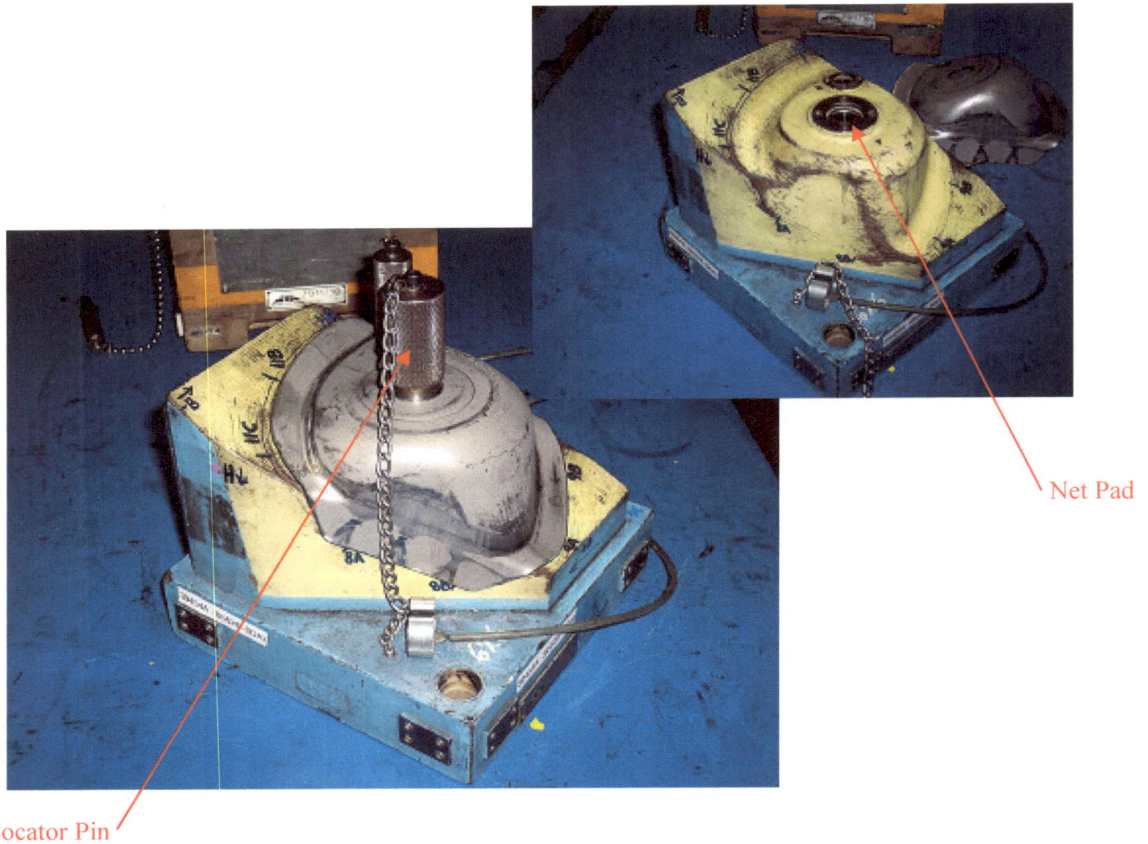

Net Pad

Locator Pin

Figure 88

In figure 88, the fixture is made from one piece of material, the fixture is cut 3mm clear of the part, so these surfaces can be feeler checked. The blocks on the side of the block are "alignment blocks" used to align the fixture to the CMM for certification of the fixture or CMM measurement of the part. Note that on this fixture, there are no clamps as the part is located via a locator pin and gravity holds the part on the net.

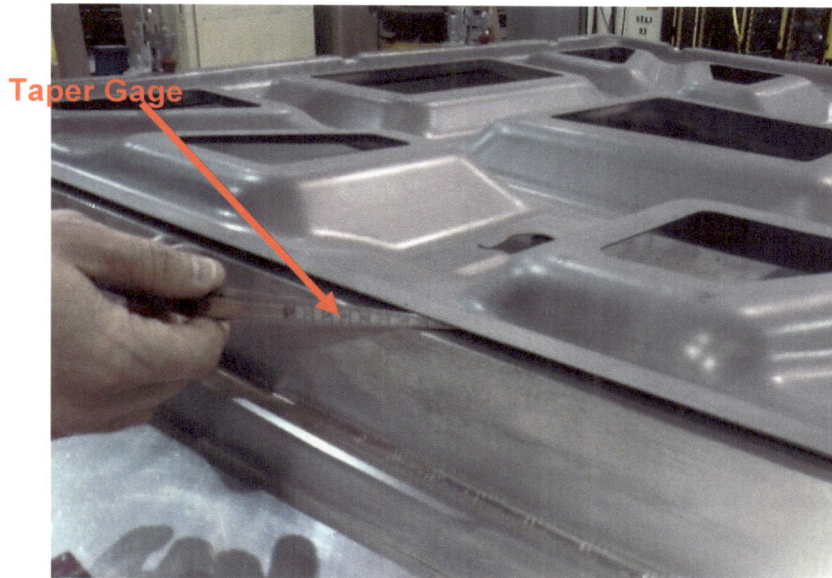

Figure 89 – Taper Gage check of the gap between the part and the fixture (this fixture has a 6.0mm nominal gap)

Figure 90 represents some of the questions that need to be considered when reviewing a check fixture for buyoff. These questions, while not all-inclusive, are designed to cover the basic functionality of the check fixture, including operator safety. Specific customers or companies have additional requirements that may also need to be captured at the time of buyoff.

Check Fixture Build Buy-off Checklist

	CHANGE LEVEL & IDENTIFICATION
1	Is fixture built to latest engineering level? If no, explain.
2	Is fixture stamped with engineering change level built to, revision date, part name, part number, customer and fixture build source?
3	Customer owned tooling, gaging and equipment is properly identified
4	Are details properly identified (part numbers, detail numbers, etc.)?
5	Are style change areas or unique modules clearly identified with part number, detail and or gage numbers?

	DATUMS
6	Are all datums (2-way, 4-way locators and net blocks) identified?
7	Are all datums (2-way, 4-way locators and net blocks) built to print specifications (MMC, RFS, etc.)?

	CLAMPS
8	Do pressure feet on clamps fall directly on the net surface area? (Remove part to check)
9	Do all clamps contact part on their correct arc of vector?
10	Are clamps adjusted to proper material thickness?
11	Are clamps numbered with clamping sequence used from acceptable Gage R?

	BASE
12	Is fixture base large enough to capture all details in open position?

	PINS
13	Are check pins, scribe pins and go/no go check pins assigned a color, letter or numeric value matching the hole chart?
14	Are all pin checks identified with correct sizes from the GD & T ? (Check to the GD & T)

	HOLES
15	Are all holes accounted for that are listed on the GD & T? (Check to the GD & T) List Quantity_____
16	Do all blind hole checks have scribe pin capability?

	FEELERS, GAP / FLUSH
17	Are gap & flush check areas labeled on fixture?
18	Are feeler checks able to check all form as per design?

	DETAILS & HAND APPLY GAGES
19	Are hand applied and interchangeable units stamped with part and or gage no's and stored as per gage design?
20	Are loose details permanently attached to fixture where possible?

	SPC
21	Are all SPC points labeled on fixture with minimum 2 characters from the GD & T (characteristic label)?
22	Do LMI probes, Dial Indicators, etc. fit in all SPC bushings correctly?

	GAGE R & R
23	Were tolerances used for Gage R verified to tolerances on GD&T?
24	Has fixture been checked for obstructions with part before performing Gage R?

	OPERATOR INSTRUCTIONS
25	Does O/P procedure on fixture reflect clamping sequence used from acceptable Gage R?
26	Is O/P procedure for fixture clear and consise?
27	Does the O/P sheets capture all removeable, locking, retracting, etc. items correctly?

Figure 90

Another method to verify, not only the part, but the process, is scanning. Here a part may be placed on a fixture or on supports, depending if the part is rigid enough to hold its shape, without deforming, under gravity load. The part is then scanned, using multiple methods, either a laser scanner, a white light scanner or a photogrammetry system.

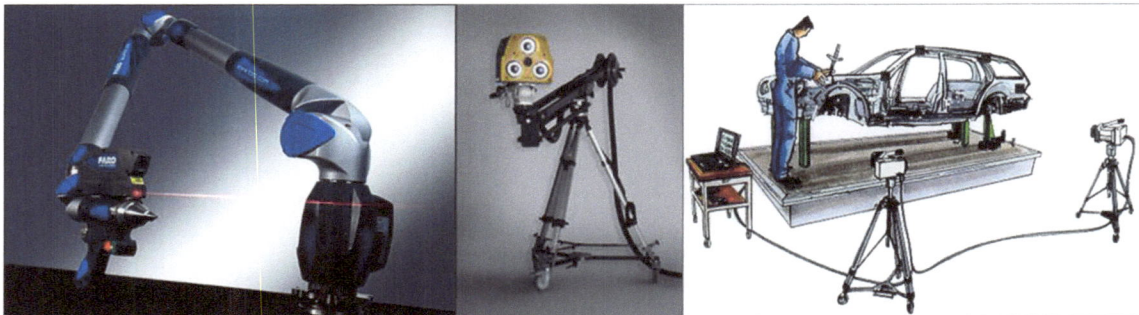

Manual Laser Scan System (Faro Technologies)

Manual White Light Scan System (Cognitens)

Manual Photogrammetry System (Geodetics)

Once the part is scanned, it can be compared to the part CAD model for acceptance. This can also be done with dies and assembly tools, as noted in Chapter 4.

Below, are two examples of data outputs from scans, aligned the same, but providing different information, due primarily to what side of material the scanning system is comparing the scan data to the CAD.

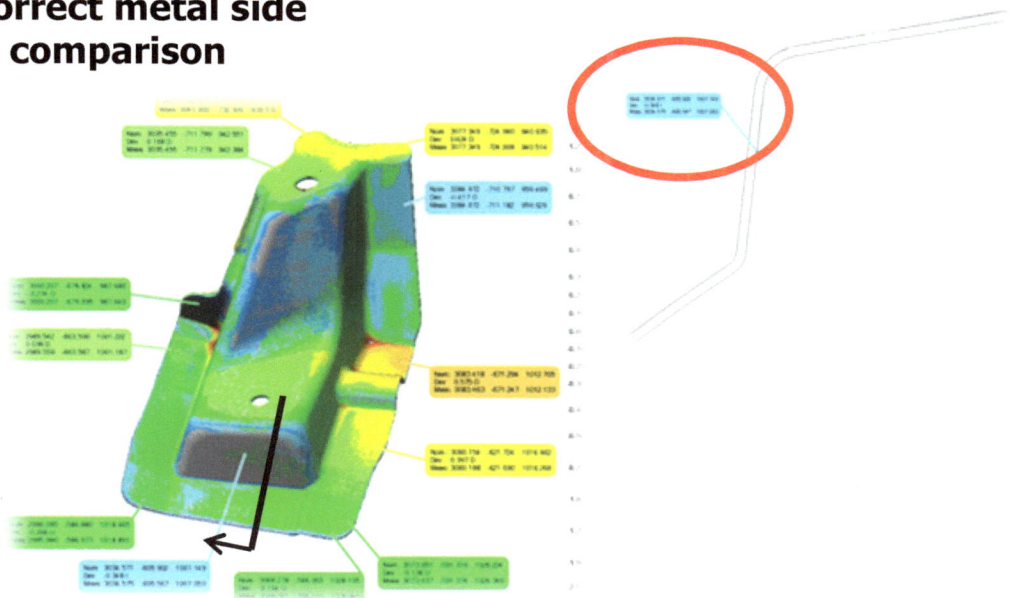

Incorrect metal side comparison

Figure 91 – Scan data compared to incorrect material side

Correct metal side comparison

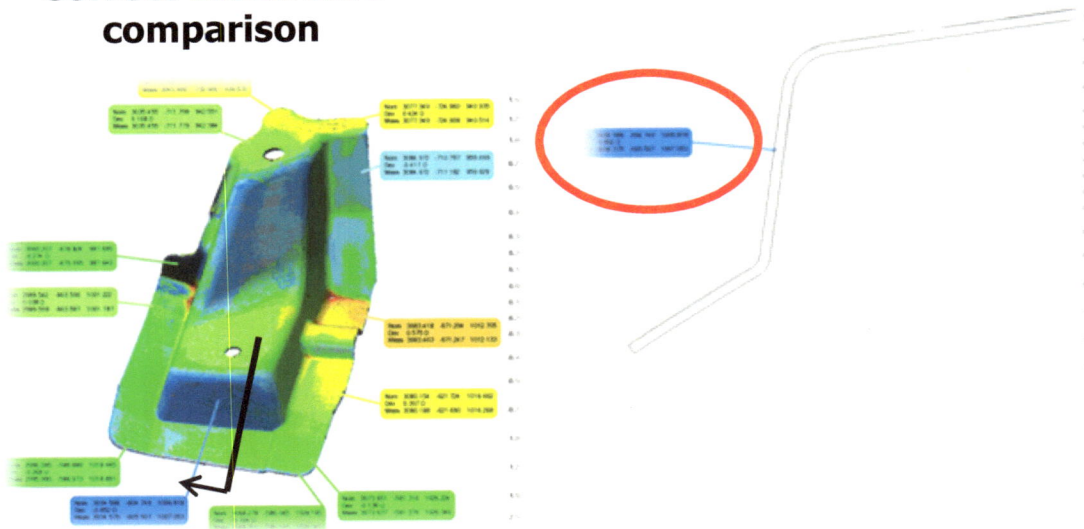

Figure 92 – Scan data compared to correct side of material

In these examples (figures 91 & 92), the scan data is exactly the same, so are the alignment points. The difference is that the measurement system compared the scan data to one side of the material in one output and to another side of material in another output. This may be caused as the scan information resulted somewhere between the two "material" thickness boundaries or a different vector associated to this particular surface. In either case, it is important to understand the data before making any decisions on what to do next.

In this case, the surface in question was not critical to the overall part/assembly so no change was necessary. However, if this was a critical surface, without understanding the data provided thoroughly, a change in the parts' die or mold could have been made in the wrong direction or amount resulting in cost and timing issues and next operation build issues.

Functional Build (or Plate Build) methods

A "functional" build is typically a single part build where details are placed together in assembly sequence on a fixture; then mechanically fastened together manually either by screws, tack welds, or other means in the same sequence as in the assembly process.

The over-riding purpose of this build is to determine how mating surfaces, from one detail compare to the next, and the effects on the build as they are fastened together.

In order to get the most beneficial data from this exercise, the following steps must be in place:

1. All details for the functional build must show repeatability:
   While only one set of details are used for this event, they must be representative of the production process used to manufacture them. They also must be repeatable. If too much variation exists in the detail process, then the functional build will only show a portion of what may be present in the assembly.
2. All functional build fixtures must be certified.
3. All functional build fixtures must meet the assembly process locating scheme.
4. A complete process flow of the assembly including fastening methods and sequence must be available at the start of the functional build.
5. Functional build should be performed on a CMM. This will allow additional data gathering through each phase of the assembly buildup, including measurements after each fastener is implemented, etc.

While there are many advantages to the functional build including:
   A. Understanding of the detail's effects on the assembly
   B. Understanding of the fastening sequence on the assembly
   C. Understanding the effects of the locating scheme on the assembly.

It cannot predict the influences of the following items that can be witnessed in the assembly process itself:
   A. Fastener heat effects, from welding
   B. Simultaneous fasteners
   C. Effects of transferring mechanism (i.e. robot transfer, etc.)

Figure 93

An example of a functional build fixture is shown in figure 93 above. This one fixture represents the complete rear floor assembly process. Shown in the picture is the start of the build up process, where the lower rails and upper rails are mated together (i.e. left hand and right hand).

Items in RED are the various clamps used for the assembly process and are sequenced according to the actual production process. Once the initial build-ups are made, various clamps can be removed or swung out of the way if they are not part of the next assembly process.

Figure 94

Figure 94 shows the front and rear lower rail details mating together prior to fastening. Here, the two details are located on the fixture by their respective datums.

The mating surface gaps, highlighted in the ovals above, indicate areas where stress may be induced into the build once the fasteners are implemented and the clamps are released (opened). The design of this product has all mating surfaces in a line-to-line condition. So, theoretically, all surfaces should sit on top of each other with no gap.

Design, being a theoretical nominal, will not account for all tolerance related conditions possible. In the figure above, it is possible that both details, individually meet the tolerance requirements; but at opposite sides of nominal so when placed together, the details will not mate properly.

Figure 95

Figure 95 represents the front and rear lower rails are fastened together. Here, the locators for the rear lower rail are removed (i.e. dropped down) for the next assembly process.

In this particular functional build, shown above, screws with nuts are used to simulate welds and assembly fused locations. Here, seven "welds" hold the rear rail to the front rail.

The scenario above, where multiple surfaces are fastened together, indicates the importance to understand the effects of various weld sequences. For this reason, numerous measurements can be taken at various steps in the fastening process. Some of these sequences may be:

Trial Number	Weld Sequence
1	1,2,3,4,5,6,7
2	4,5,3,6,2,7,1
3	4,5,6,7,3,1,2

While these are not all of the sequences that may be tried, it is important to understand for this build what geometry directions are most important, and base the sequence off of that. In this instance, the z direction or up/down is most important (welds 4 and 5) and followed by the in/out or y direction (all remaining welds). It is also important to understand that y welds 3 and 6 are on surfaces with the most structure. Since they are next to the bend radius, there will be little deformation of the part at this location and should be welded next.

Virtual Assembly Build methods

Another method of understanding the fit up of detail parts to each other is to do a virtual build. This is done by scanning detail parts, then building up the assembly in a CAD environment using one part as the master, then bringing the scan of the next part to it.

Benefits of this method include:
1. Elimination of fixture costs associated with a plate build
2. Parts can be aligned to each other in a variety of methods including by:
    a. Each parts locating features
    b. Each parts mating features to the next part

Example:
Scan of detail part "A" relative to it's locators

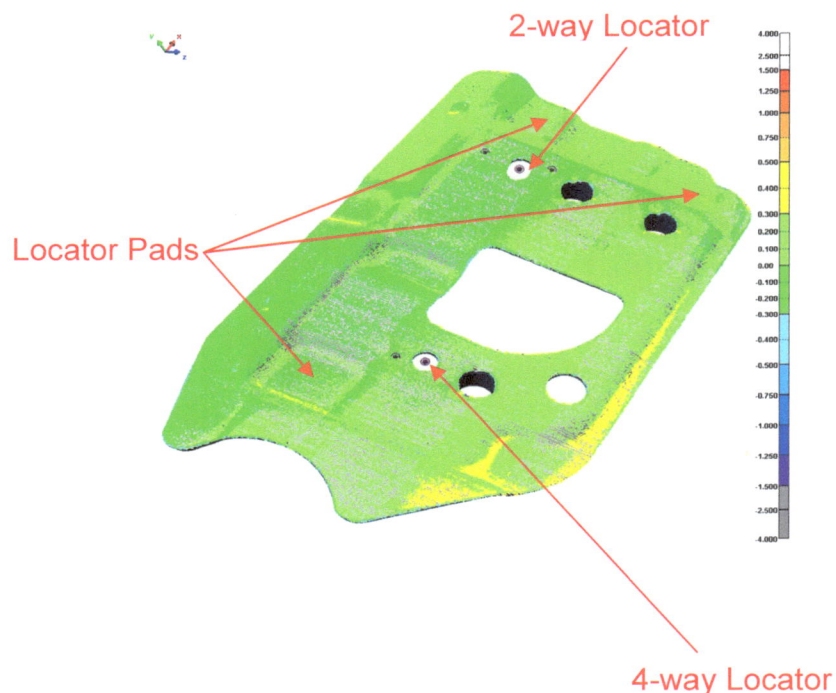

Figure 96

While this figure shows the deviations of the part relative to the locators designated or aligned to, in cases where more than three planar points exist, it is also important to understand the deviation of the locators relative to the best fit plane.

Scan of detail part "B" relative to it's locators

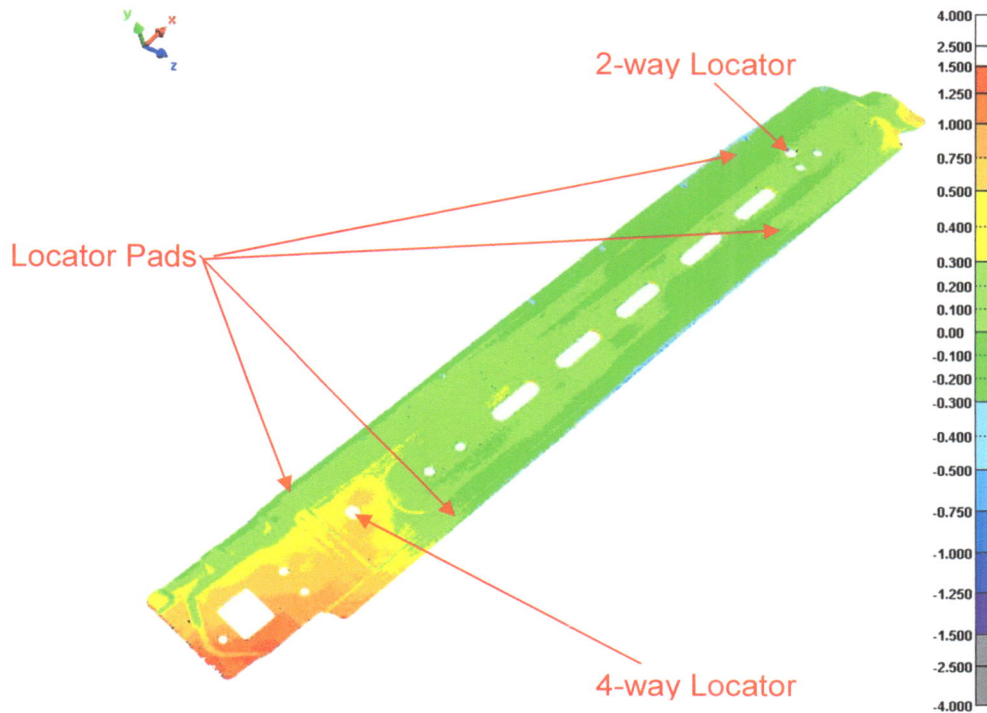

Figure 97

From this, detail "B" scan is converted to a CAD file, which will then be used as the "master" for detail "A" to be compared to. In figure 98, detail "A" is aligned to detail "B" via detail "A's" datums. Here, since the datums are not on the mating weld flanges, the resulting "virtual build" shows a gap between both parts. In both cases, part "B" as the master, is set up or located via it's locating scheme.

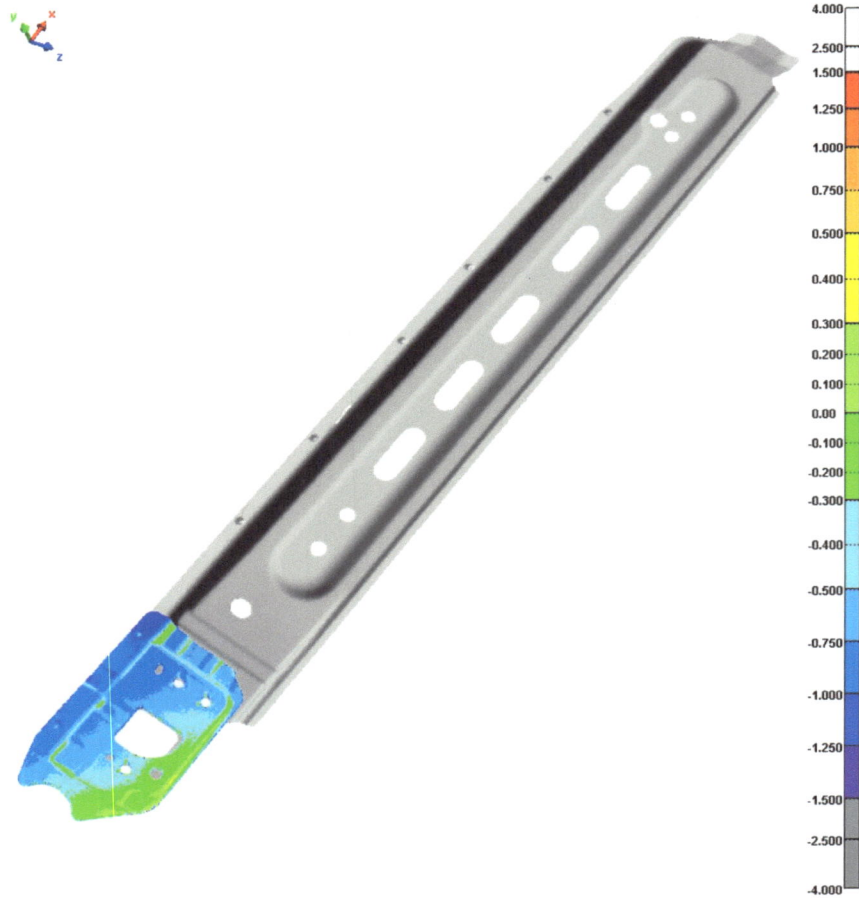

Figure 98 – Virtual build aligning to part "A" datums

Note that while this virtual build result shows a "gap" of up to approx. 1.0mm, if the part was not scanned on the correct side or the proper material compensation (thickness) wasn't applied to the appropriate scan, the result may be incorrect.

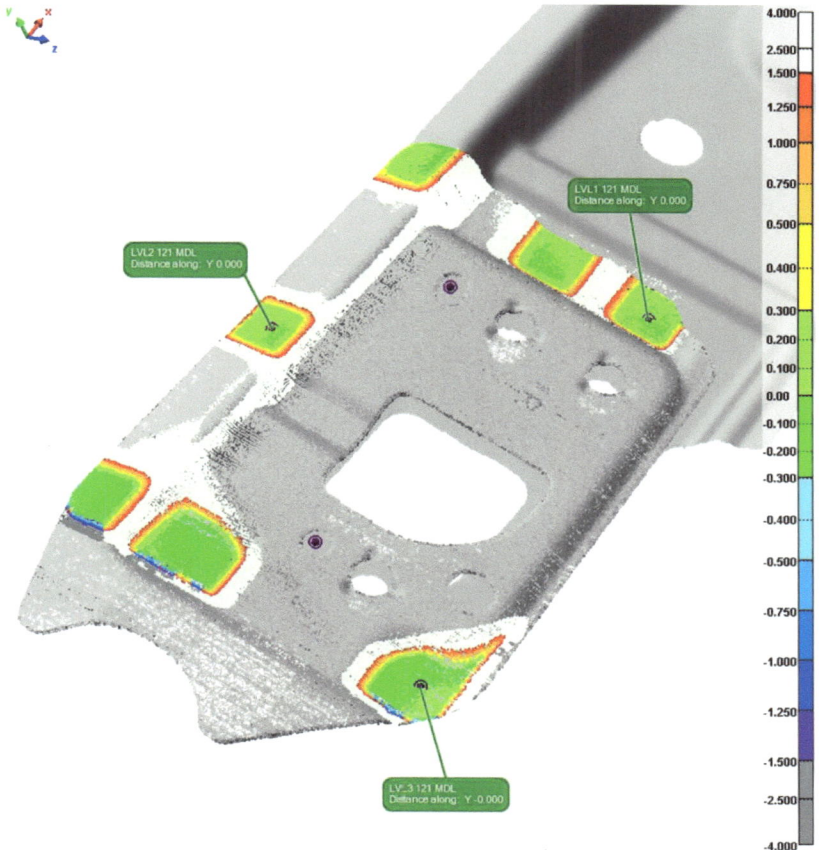

Figure 99 – Virtual build alignment of mating surfaces

In figure 99, alignment of detail part "A" to "B" is via the mating surfaces, this then shows that the rest of the part falls outside of the color scale (>1.50mm deviation)

Chapter 9

"C"

PPAP (Production Part Approval Procedure) FLOW CHART

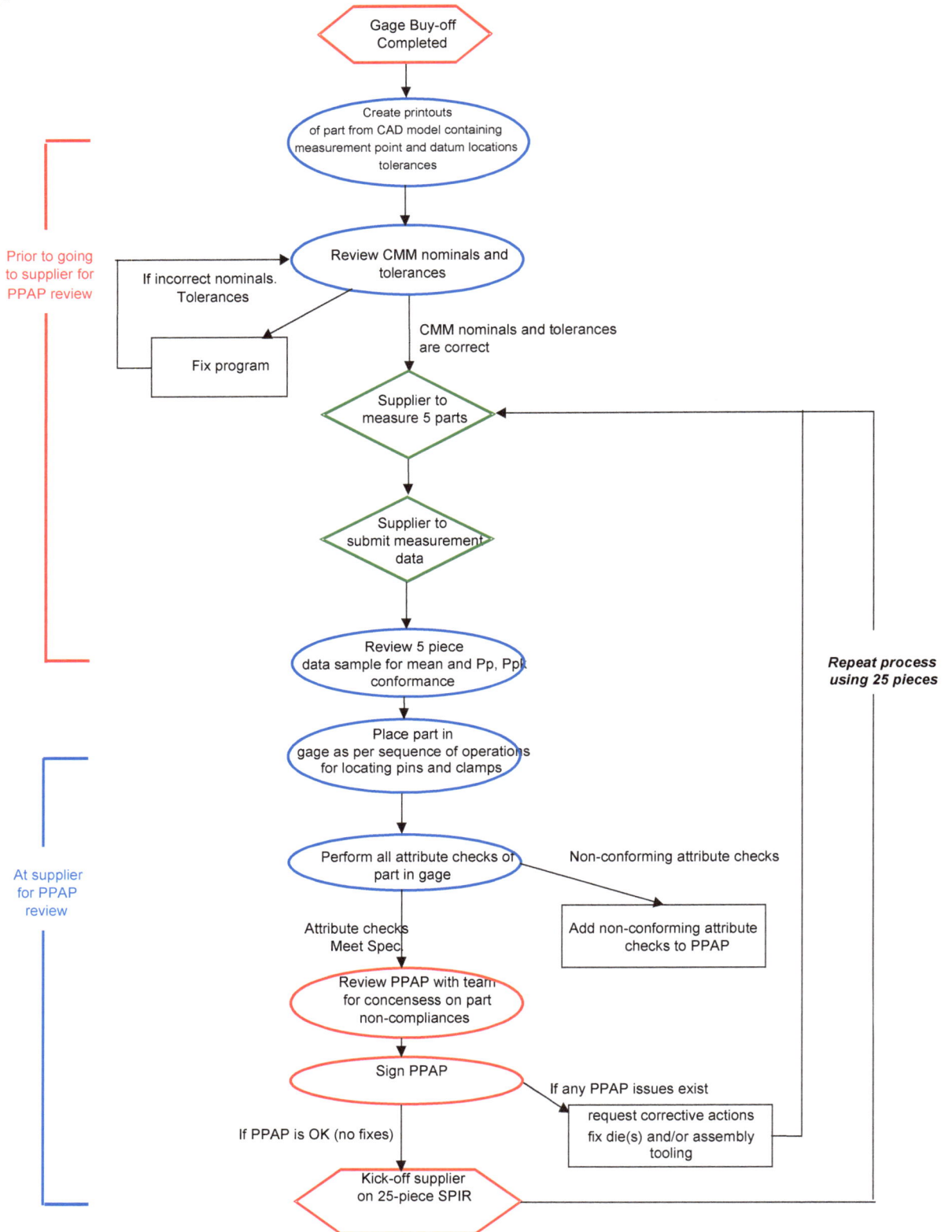

```
                          Gage Buy-off
                          Completed

                          Create printouts
                  of part from CAD model containing
                  measurement point and datum locations
                          tolerances

                          Review CMM nominals and
                          tolerances
```

Prior to going to supplier for PPAP review

If incorrect nominals. Tolerances

Fix program

CMM nominals and tolerances are correct

Supplier to measure 5 parts

Supplier to submit measurement data

Review 5 piece data sample for mean and Pp, Ppk conformance

Place part in gage as per sequence of operations for locating pins and clamps

At supplier for PPAP review

Perform all attribute checks of part in gage

Non-conforming attribute checks

Add non-conforming attribute checks to PPAP

Attribute checks Meet Spec.

Review PPAP with team for concensess on part non-compliances

Sign PPAP

If any PPAP issues exist

request corrective actions fix die(s) and/or assembly tooling

If PPAP is OK (no fixes)

Kick-off supplier on 25-piece SPIR

Repeat process using 25 pieces

PPAP (Production Part Approval Process) Requirements

Production part approval is usually required prior to the first production shipment of the product. Production Part Approval Process (PPAP) is a North American Automotive OEM (NAAOEM) standardized method for obtaining buyoff of components for production run. While the standard is a NAAOEM one, a lot of the items listed are used by various other OEMs.

To effectively PPAP a part or assembly with minimum scrap or downtime between production runs, a PPAP timeline should consist of a 5-25 process. That is, a batch process that allows for an initial 5 parts to be run and measured. Once this batch is measured, the team can assess it's quality and either continue with the run or recommend part changes. If the run is continued, then 25 more parts are then taken from the overall production run for measurement. This set of 25 will be combined with the 5 already measured to create a 30-piece study. If part changes are recommended, then the process starts all over.

It is also important to understand part "areas of concern". These are areas of the part that were identified on "build issues" on previous prototype/pre-production build events. Focusing attention on these items will help address issues prior to the parts reaching the measurement area. It can also reduce production equipment downtime or excessive scrap to be made.

It is recommended that part issues be identified based on the following priorities:
1. Functional (Build issue)/Customer issues
2. Functional/Non-customer issues
3. Dimensional/non-functional issues

Additionally, it is important at this phase that all production methods are followed. In a stamping process: ensure the dies are in the homeline press, the die setup process is followed, transfer fingers are functional, and so forth. Assembly processes should be run in the "automatic" mode. Following these rules will ensure that data generated from this run will reflect normal production methods of a single run. A single run, because a PPAP pass, will almost never include multiple line setups, multiple incoming component runs, etc.

The reason for measuring an initial 5 parts is to determine if all features of the part are present (i.e. holes, flanges, etc.) and the dimensional data, by reviewing the mean and range, is acceptable.

A general rule is that the "mean" should be no greater than 20% of the tolerance away from nominal; with a "range" less than 40% of the tolerance to hope to achieve NAAOEM PPAP requirements (see page 94 for the AIAG requirement). Another, more simpliar approach might be to use a Ppk value that is at the high end of the confidence band for these 5 parts. (This method is explained further on pages 151 and 152). These guidelines will help in addressing areas well before more parts are run and the entire lot scrapped.

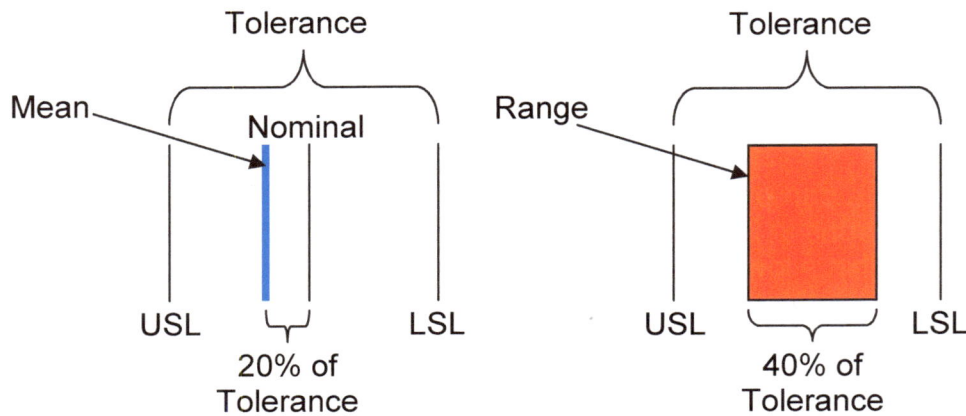

Figure 100

Combining the two indicators, we will have a better understanding if the parts will end up inside of the tolerance. Please keep in mind, however, that AIAG currently requires an initial 1.67 Ppk at PPAP to achieve a production life run of 1.33 Ppk.

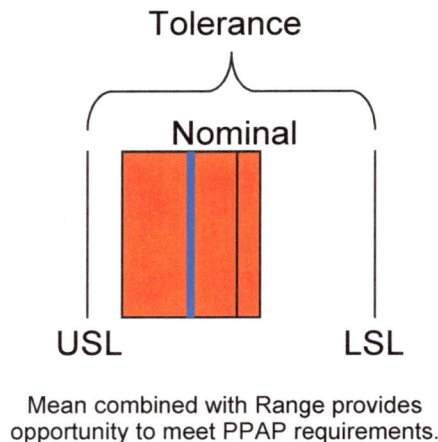

Mean combined with Range provides
opportunity to meet PPAP requirements.

Figure 101

For multiple geometry set stations, it is important to determine the capability of each station by itself and as a total. Another important factor to consider is that a total capability using the standard 6-sigma calculation for normal distribution may not be accurate as each station; it may have a mean deviation to itself; but when combining the data from both stations, provide a 6-sigma value larger than what may actually be seen in normal production. See example figure 99.

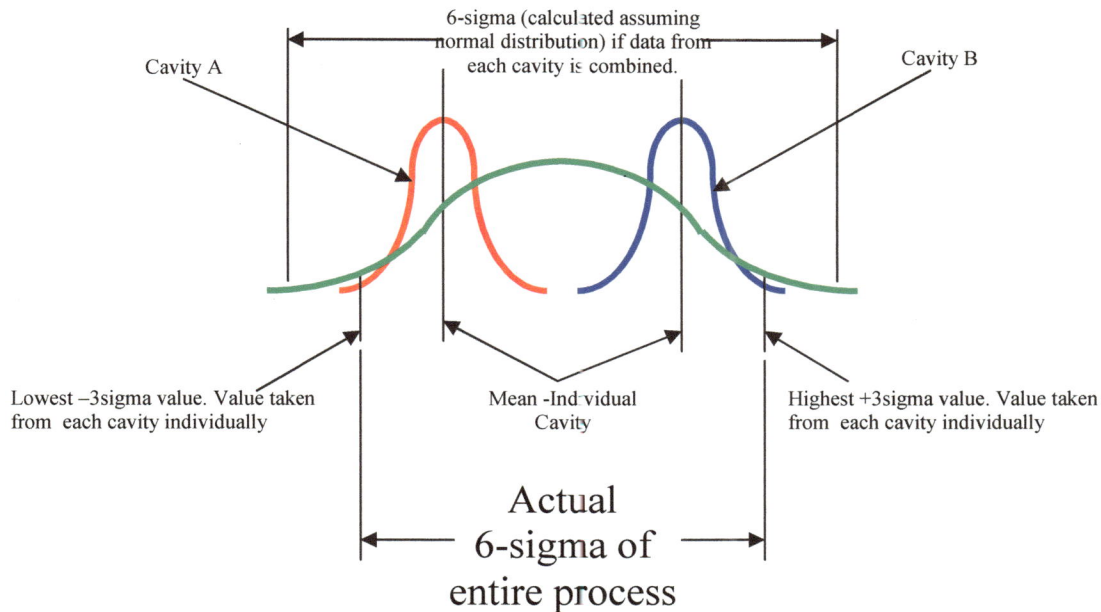

Figure 102

In the above figure, the part measured is part of a two-out process. Stated another way, that the die process produces two parts for each stroke or cycle. Cavity A represents one part and cavity B the other part from this process. Here, Cavity A has a mean deviation that is different than cavity B. By combining the two cavities together, the resulting bell curve or 6-sigma distribution would be the "green" curve, which is not the sum of the individual cavities worst case 6-sigma values (see figure 100).

	Pp	Cr	Ppk	High Limit	Low Limit	Mean	Range	Min.	Max.	Sigma	+3 Sigma	-3 Sigma
CAVITY A, Pt 1	3.481	0.287	2.397	0.70	-0.70	0.218	0.230	0.100	0.330	0.0670	0.4191	0.0169
CAVITY B, Pt 1	3.585	0.279	3.170	0.70	-0.70	-0.081	0.230	-0.200	0.030	0.0651	0.1143	-0.2763
Both Cavities	1.420	0.704	0.993	0.70	-0.70	0.211	0.530	-0.200	0.330	0.1644	0.7037	-0.2826

Figure 103

In this case, it is important to provide the maximum and minimum 6-sigma from each cavity. An example of a different mean deviation, range, etc. from each cavity and the resulting overall part capability (both cavities) is shown in figure 104.

	Pp	Cr	Ppk	High Limit	Low Limit	Mean	Range	Min.	Max.	Sigma	+ 3 Sigma	- 3 Sigma
CAVITY A, Pt 1	3.481	0.287	2.397	0.70	-0.70	0.218	0.230	0.100	0.330	0.0670	0.4191	0.0169
CAVITY B, Pt 1	3.585	0.279	3.170	0.70	-0.70	-0.081	0.230	-0.200	0.030	0.0651	0.1143	-0.2763
Both Cavities	2.013	0.497	1.816	0.70	-0.70	0.069	0.530	-0.200	0.330	0.1159	0.4191	-0.2763

Figure 104

This will then provide a truer distribution of what this process will produce, it also reduces the possibility of reworking a process that, if both cavities were combined into one calculation (same as figure 100), wouldn't pass typical PPAP criteria.

Tolerance Concessions

When a part provides a good quality next assembly, but the detail part is out of tolerance or the appropriate build issues areas of the part have been corrected and remaining "out-of-tolerance" areas do not effect the next build phase, tolerance concessions may be the next approach to achieve component buyoff. When approaching potential tolerance concessions, we must understand how the part will be continuously monitored over its production life. We should address the following questions: Is the part to be measured on a full attribute check fixture? Will it be monitored only on a CMM? It is important to understand these questions so that the tolerance concessions recommended are not only based on the current product quality produced; but also address how it will be continually verified throughout the production lifecycle.

In the example below, we show an example of two different methods of calculating the tolerance concession:

     1. Tolerances based on specific measurement point locations. If the only way to validate part acceptability is via a coordinate measurement machine, then a point-by-point tolerance may be valid. Although the tolerance feature control frame is based on geometry of the surface or feature.

     2. Tolerances based on buyoff criteria from a coordinate measurement machine but the production validation methods are from an attribute fixture. Then, looking at it from a perspective of how and who will be validating the part is necessary.

In figures 105 through 109, is an example of how a tolerance by point ID may not be best in attribute validation of the part during production and a recommendation.

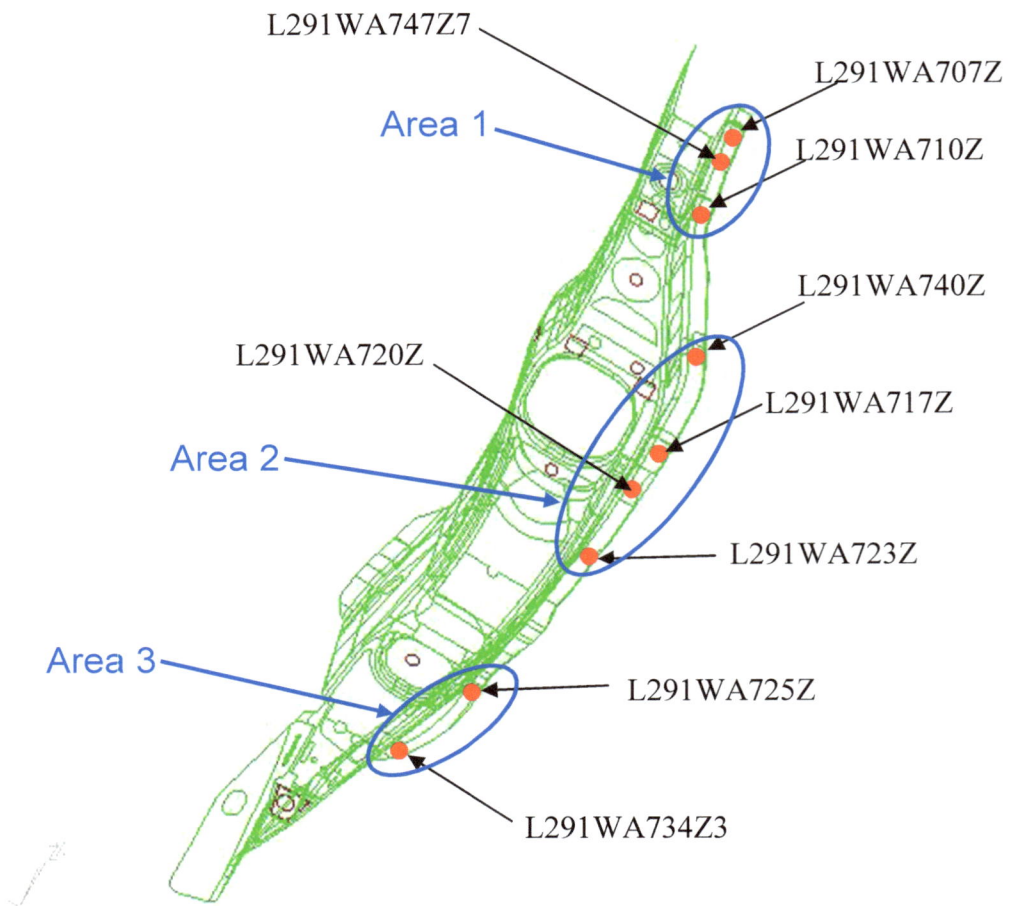

Figure 105

Original Tolerances						
	Pp	Ppk	High Limit	Low Limit	Mean	Range
L291WA707Z.	3.03	-0.01	0.7	-0.7	0.702	0.300
L291WA710Z.	4.77	3.01	0.7	-0.7	-0.257	0.184
L291WA717Z.	4.56	-0.81	0.7	-0.7	-0.825	0.207
L291WA740Z.	5.15	5.09	0.7	-0.7	0.008	0.185
L291WA747Z7.	3.69	3.20	0.7	-0.7	-0.093	0.243
L291WB723Z.	4.67	2.28	0.7	-0.7	-0.358	0.218
L291WB725Z.	4.88	4.81	0.7	-0.7	-0.011	0.210
L291WA734Z3.	4.04	2.12	0.7	-0.7	-0.332	0.308

Figure 106

Statistical data generated from a 30-piece sample

Tolerance Concessions - Based on Specific Point Locations						
	Pp	Ppk	High Limit	Low Limit	Mean	Range
L291WA707Z.	3.03	1.72	1.10	-0.30	0.702	0.300
L291WA710Z.	4.77	3.01	0.70	-0.70	-0.257	0.184
L291WA717Z.	4.56	1.80	0.30	-1.10	-0.825	0.207
L291WA734Z3.	4.04	2.12	0.70	-0.70	-0.332	0.308
L291WA740Z.	5.15	5.09	0.70	-0.70	0.008	0.185
L291WA747Z7.	3.69	3.20	0.70	-0.70	-0.093	0.243
L291WB723Z.	4.67	2.28	0.70	-0.70	-0.358	0.218
L291WB725Z.	4.88	4.81	0.70	-0.70	-0.011	0.210

Figure 107
Yellow highlighted area designates tolerance changes based on this method

In figure 107, the tolerance concessions in yellow are applied to only the specific point locations. While the GD&T feature control frame typically uses the "profile of surface" symbol, the tolerance in this instance will be part of the point ID callout in the product CAD file.

Tolerance Concessions - Based on Part Features						
	Pp	Ppk	High Limit	Low Limit	Mean	Range
L291WA707Z.	3.46	1.72	1.10	-0.50	0.702	0.300
L291WA710Z.	5.45	1.65	1.10	-0.50	-0.257	0.184
L291WA717Z.	4.89	1.80	0.40	-1.10	-0.825	0.207
L291WA734Z3.	4.33	4.23	0.40	-1.10	-0.332	0.308
L291WA740Z.	5.52	2.88	0.40	-1.10	0.008	0.185
L291WA747Z7.	3.96	2.60	0.40	-1.10	-0.093	0.243
L291WB723Z.	4.67	2.28	0.70	-0.70	-0.358	0.218
L291WB725Z.	4.88	4.81	0.70	-0.70	-0.011	0.210

Figure 108
Statistical data generated from a 30-piece sample
Yellow and Blue highlighted areas designate tolerance changes based on this method

In the tolerance concessions, figure 108, we divided up the "z" (up/down) surface into three areas by applying a tolerance for each area. For an attribute check, individual "feeler" templates would be needed for each of these areas. If we chose to get down to one tolerance, we would end up with a +1.00/-1.00mm tolerance across the entire surface. In either case, the feature control frame(s) will need to be updated to these tolerances. If the feature control frame can be attached in CAD, to the surface ID, then we are done. If not, then start and stop locations may need to be identified for the feature control frames.

When planning PPAP (Production Part Approval Process) timelines for components, both detail and assembly, never assume the first run will meet all requirements. In general, it is recommended that depending on the complexity of the parts, one should anticipate a minimum of two production setups. You should anticipate some changes in between, before making the final PPAP run.

A PPAP timeline should include all necessary processes for creating the component and processes for measuring the component, including any destruct testing where required.

Figure 109 below depicts a PPAP readiness matrix, which identifies and tracks all the steps required for a component to be ready for PPAP. This matrix is only an example. Some types of components may require additional PPAP information (i.e. sample parts, etc.) or less depending on customer criteria.

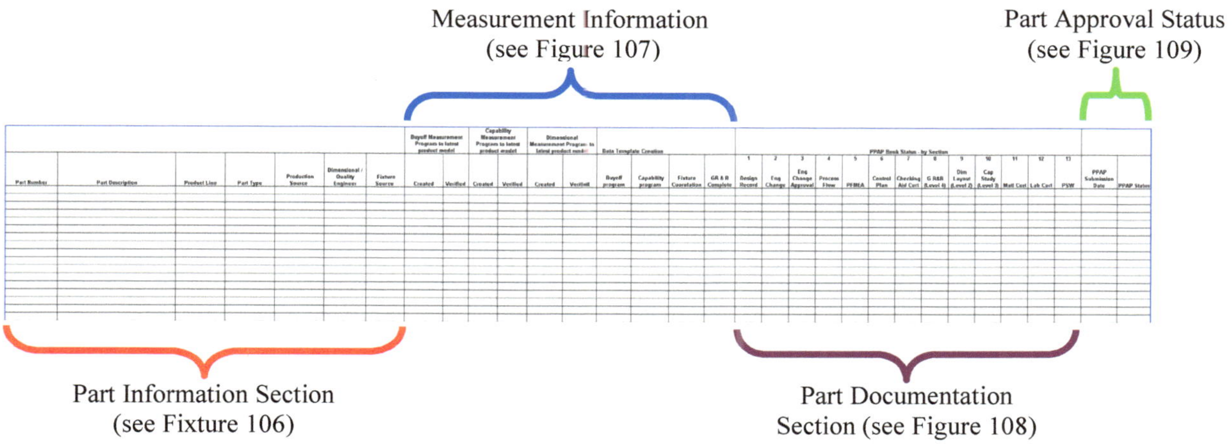

Measurement Information
(see Figure 107)

Part Approval Status
(see Figure 109)

Part Information Section
(see Fixture 106)

Part Documentation
Section (see Figure 108)

Figure 109

Part Information Section

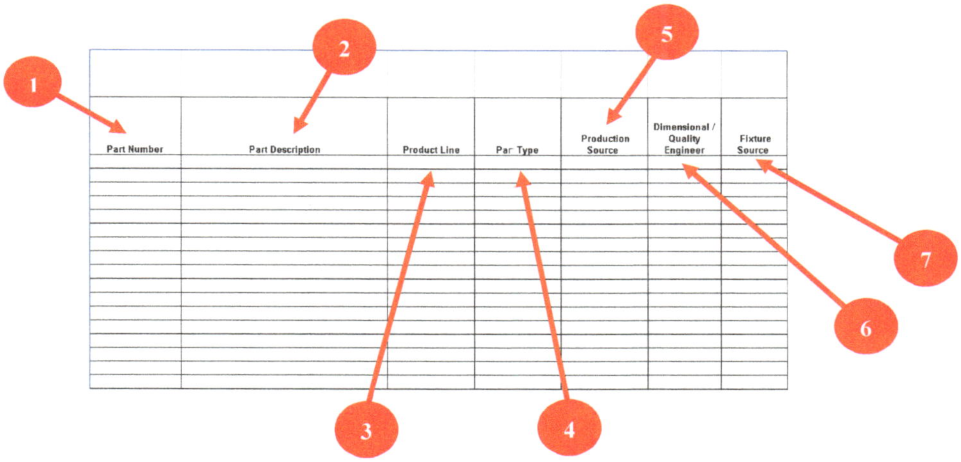

Figure 110

The part information section contains all the necessary information to track the status of the part including:
1. Part Number
2. Part Description
3. Product Line
4. Part Type (i.e. component, sub-assembly, etc.)
5. Production source (i.e. where the part is produced)
6. Dimensional/Quality Engineer responsible for the part
7. Part Fixture source (if necessary)

Measurement Information Section

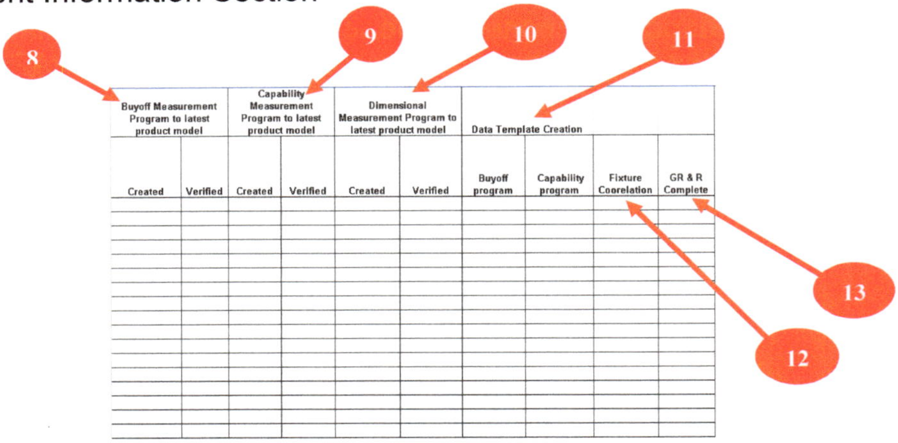

Figure 111

The measurement information section contains all the necessary information to track the status of the part including:

8. Buyoff measurement program creation and verification
9. Capability measurement program creation and verification
10. Dimensional measurement program creation and verification
11. Data template creation
12. Fixture Correlation – correlates the fixture results to the CMM results so both are equal.
13. Gauge R & R status

Part Documentation Section

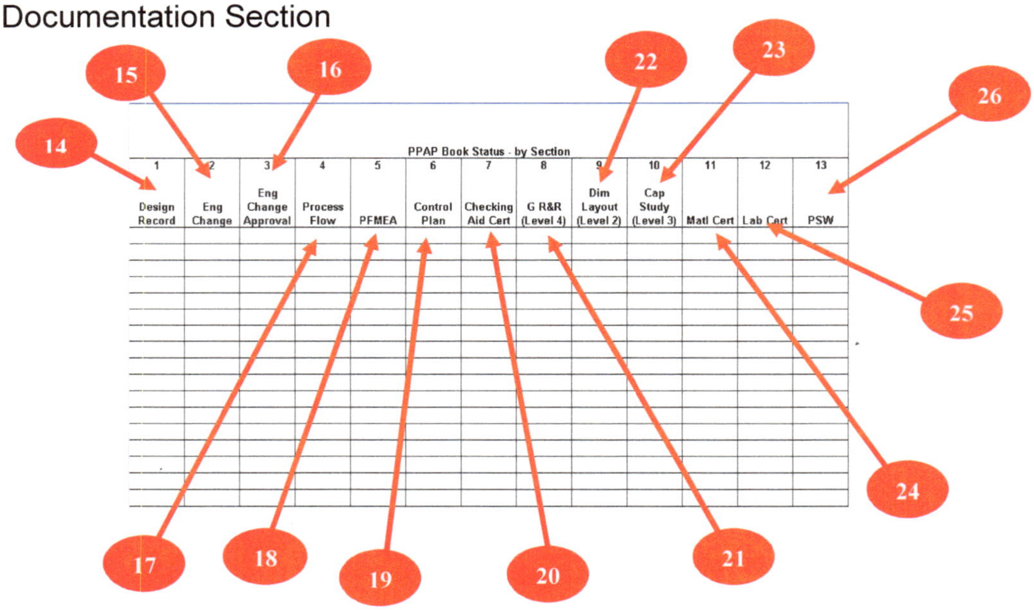

Figure 112

The part documentation section contains all the necessary documentation for each component needed to PPAP. The documentation tracks the status of the part including, but not limited to:

14. Part Design Record
15. Part Engineering Change Level
16. Part Engineering Change Approval
17. Process Flow Chart – flow chart of the process used to create the component.
18. PFMEA
19. Control Plan
20. Checking Aid Certification – copy of the certification of all measurement devices used to measure the component (i.e. fixture, CMM, etc.)
21. GR&R results
22. Dimensional Layout results
23. Capability Study results
24. Material certification – certification records of the material(s) used to make the part.
25. Lab certification
26. Part Submission Warrant (PSW)

Part Approval Status Section

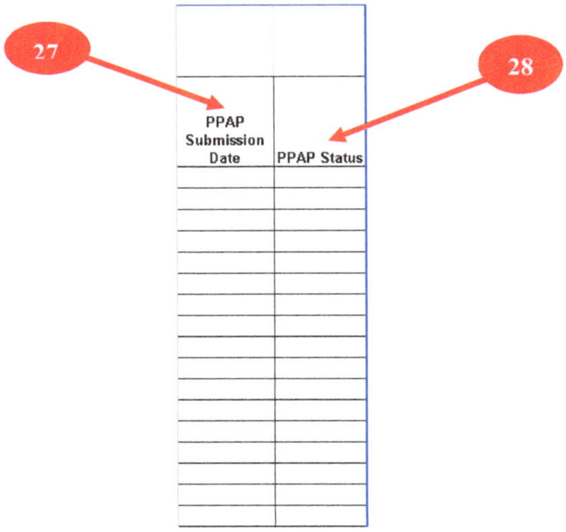

Figure 113

The part documentation section contains all the necessary documentation for each component needed to PPAP. The documentation tracks the status of the part including:

        27. PPAP Submission date – date the PPAP package was submitted for approval.
        28. PPAP status – identifies if the part was approved or rejected.

PPAP is also dependent on the standing of the production source. As identified in the AIAG manual, there are five levels of PPAP. The production source is assigned a certain level by its customer depending on previous quality, delivery, and productivity requirements.

We have so far discussed Pp, Ppk, Cp, Cpk, PPM, mean, range, etc. In this next section, we will attempt to familiarize these statistical indices with the meaning and equation for generating them.

Process Performance Index (Pp, Ppk) - Total process variation, including common and special causes. The process is not necessarily stable or in-control. The index is calculated by:

$$Pp = \frac{(USL - LSL)}{6\sigma_s}$$

$$Ppk = \frac{(USL - x)}{3\sigma_s} \text{ or } \frac{(x - LSL)}{3\sigma_s} \text{ ,Whichever is lesser.}$$

Where:

USL = Upper Specification Limit
LSL = Lower Specification Limit
σ_s = The estimate of the standard deviation of a process using the sample standard deviation of a set of individuals about the average of the set.

Pp and Ppk should be used on prototype and pilot components (pre-launch) since sample sizes are typically small (less than 125) and the process unstable. A minimum of 25 samples should be measured to insure an acceptable level of confidence.

Process Capability Index (Cp, Cpk) - Inherent process variation, common causes only. Process must be stable and in-control. The index is calculated by:

$$Cp = \frac{(USL - LSL)}{6\sigma_{\overline{R}/d_2}}$$

$$Cpk = \frac{(USL - x)}{3\sigma_{\overline{R}/d_2}} \text{ or } \frac{(x - LSL)}{3\sigma_{\overline{R}/d_2}} \text{ , Whichever is}$$

lesser.

Where: $\sigma_{\overline{R}/d_2}$ = The estimate of the standard deviation of a stable process using the average range of sub-grouped samples taken from the process, usually within the context of control charts.

Cp and Cpk should be used to evaluate processes based on long term and significantly sized samples.

How Cpk relates to Parts per Million (PPM) out of specification.

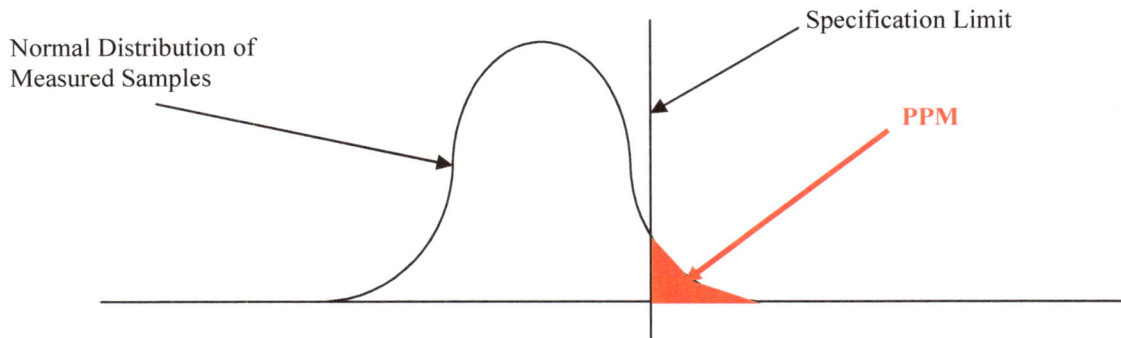

Figure 114

Out of Specification

Cpk	Percentage	PPM
0.00	50.0000%	500000.0
0.33	16.1087%	161087.1
0.50	6.6807%	66807.1
0.67	2.2216%	22215.7
1.00	0.1350%	1349.8
1.33	0.0033%	33.2
1.67	0.0000%	0.3
2.00	0.0000%	0.0

Assumptions for this table:
1. Assumes that the data accurately represents the process
2. Assumes a normal distribution that is stable
3. Uses a 6-sigma normal distribution, and
4. Lists only the percent or parts per million (PPM) out of specification in on tail of the normal distribution. (To find total, add the percentage or PPM for both Cpk (upper) and Cpk (lower).

Another factor to consider when setting the buyoff criteria for Cp and Cpk is that because this is a "snapshot" in time, long term Cp and Cpk values will be less than at initial buyoff. Generally speaking, it is assumed that +/-1-sigma will be lost over the production cycle of a product. This means, that if initial buyoff criteria is targeted at Cpk = 1.33, over the production life, the Cpk will be closer to 1.00.

Having stated this, it is important to recognize the potential amount of parts that can be out of specification, the uncontrollable factors and controllable factors that can contribute to the reduced Cpk values over the life of the product. Some of these contributors are:

- Die set variation
- Material changes (steel coil to steel coil)
- Preventative Maintenance
- Repeatability of all processes within the production of the product (i.e. how repeatable is the press tonnage, etc.)

One other item to consider is the confidence interval for the mean and variation. A confidence interval is a tolerance of sorts for the reported mean and variation based on the measured sample size. The larger the sample sizes, the smaller the confidence interval. Stated another way, the greater the trust or certainty in the actual value reported.

Confidence Intervals for Mean
Continuous Data - Large Samples
Assumes a normal distribution to calculate the confidence interval for the mean.

$$\overline{X} + Z_{\frac{\alpha}{2}} \sqrt{\frac{\sigma}{n}}$$

Where: $\overline{X}$ = the sample average
σ = the population standard deviation
$Z_{\frac{\alpha}{2}}$ = the normal distribution value for a desired confidence level

The desired confidence level means how certain you want to be in the number and range reported. Do you want to be 90%, 95% confident? In most cases, the desired level is 95% confidence.

Continuous Data – Small Samples
Below calculation is used where relatively small samples (<30) is used, then the "t" distribution must be used.

$$\overline{X} + t_{\frac{\alpha}{2}} \sqrt{\frac{S}{n}}$$

Where: $\overline{X}$ = the sample average
S = the population standard deviation
$Z_{\frac{\alpha}{2}}$ = the t distribution value for a desired confidence level and (n-1) degrees of freedom

Confidence Intervals for Variation
The confidence intervals for the mean were symmetrical about the average. This is not true for the variance, since it is based on the Chi-Square distribution. The formula is:

$$\frac{(n-1)\,S^2}{X^2_{\frac{\alpha}{2},\,n-1}} \leq \sigma \leq \frac{(n-1)\,S^2}{X^2_{1-\frac{\alpha}{2},\,n-1}}$$

Where: n = the sample average
              S^2 = point estimate of variation

$X^2_{\frac{\alpha}{2}}$ and $X^2_{1-\frac{\alpha}{2}}$ = are the table values for (n-1) degrees of freedom

Below is an example of a confidence chart. This one indicates a 95% confidence factor, meaning that based on the amount of samples measured, and the reported Cp value, the actual Cp value will lie between the two lines of like color. In this chart:

The reported Cp =1.00 and the possible actual capabilities are in green
The reported Cp =1.33 is in purple, and
The reported Cp =1.67 is in blue

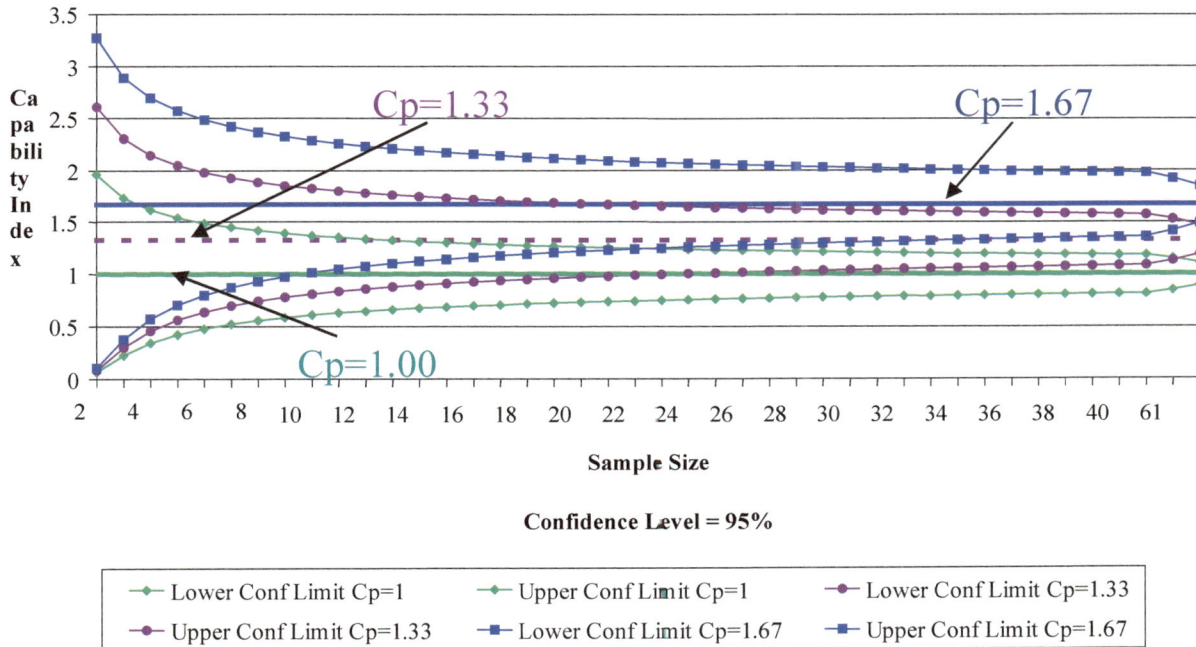

Figure 115

In example A (see figure 116), if we measured 5 parts and had a reported Cp=1.33. The actual process, based on the confidence level, could have a Cp anywhere from 0.45 to 2.15. By acquiring more samples, the amount of error in the reported Cp can be reduced.

Using the same chart, we can see that if 25 parts were measured and had a reported Cp=1.33. The process then could be anywhere between a Cp of 1.00 to 1.67

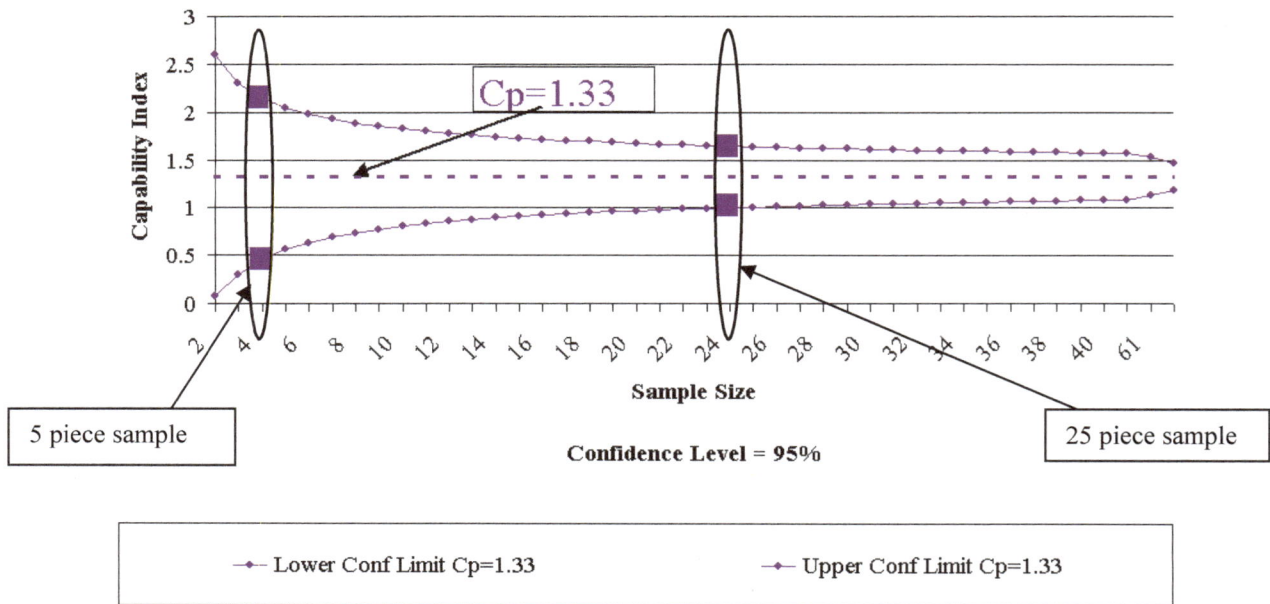

Figure 116

Using the example in figure 116, if the goal is to achieve a process that has a Cp = 1.33, then when determining whether to proceed on the buyoff process, based on the first 5 samples measured, the goal of the reported value should be around Cp = 2.15, or the upper limit of the confidence band, to reduce the potential of a part or a point falling below the 1.33 Cp target after more samples are measured.

Range - The difference between the high and low measured values. This is a quick check to evaluate tolerance conformance. Typically used to evaluate small sample runs especially during prototype and pilot build phases. This method is not statistically valid to evaluate process variability; but may give a "feel" for how close you are.

How Pp and Cp values are effected by GR/GR&R results – When a Pp or Cp value is reported, the number generated is from the calculations previously mentioned. One thing not previously discussed is related to where the measurement system error or GR/GR&R error falls. The reported Pp and Cp values were generated using the measurement system that a GR/GR&R was conducted on. Therefore, GR/GR&R error is the "noise" in the reported Pp and Cp values. Meaning that, the Pp and Cp values and the corresponding 6-sigma values have the measurement error "baked" into it. To further understand this situation, please refer to the figure below:

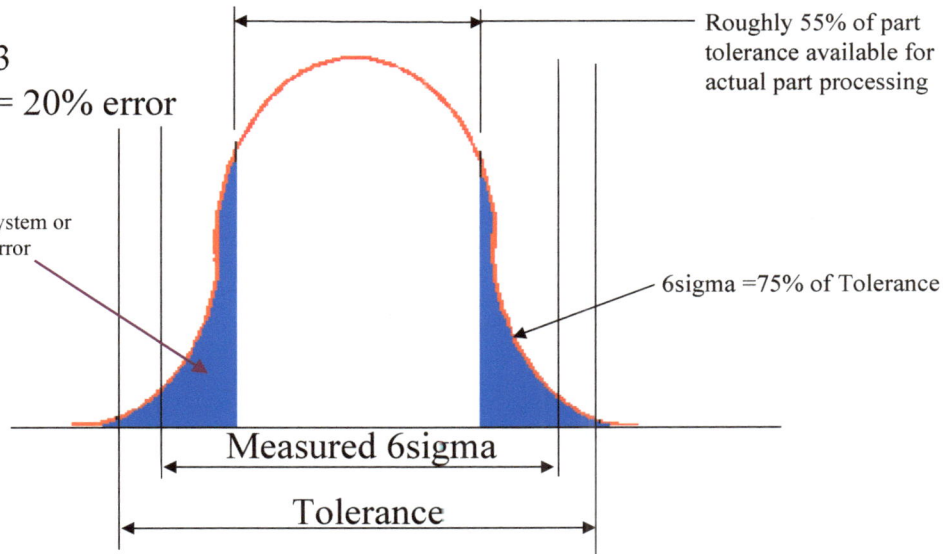

Figure 117

In the figure above, we are assuming the reported Pp value is 1.33 that indicates the process performance is using 75% of the part tolerance. The area shaded in BLUE represents the 20% gage or measurement system error at this location; thus, leaving only 55% of the part tolerance available for the actual part manufacturing or processing.

In table 118 on the next page, we will further illustrate this condition. By showing how much of the part tolerance is available, to the actual part processing, when trying to achieve various common Pp and Cp values with specific GR/GR&R measurement errors.

Pp/Cp value	Percent of Part	Measurement System Error (GR/GR&R error %)					
		5%	10%	15%	20%	25%	30%
1.00	100%	~95%	~90%	~85%	~80%	~75%	~70%
1.33	75%	~70%	~65%	~60%	~55%	~50%	~45%
1.67	60%	~55%	~50%	~45%	~40%	~35%	~30%
2.00	50%	~45%	~40%	~35%	~30%	~25%	~20%

Total Percentage of Part Tolerance available to meet Pp and Cp values.

Percentage of Tolerance remaining for part processing

Table 118

From this table, it can be determined that the higher Pp/Cp value desired to attain, the GR/GR&R (where GR and/or GR&R are calculated based on a percentage of tolerance) value must be significantly reduced so that as much tolerance as possible is available for the part processing itself.

Weld Measurement Methods

There are many different types of welds (spot welds, mig welds, laser welds, etc.) that also need to be verified for quality. Some of these welds are defined as safety critical in the DFMEA and PFMEA and control plans.

Some welds can be verified via non-destruct testing while others rely on destructive testing. The difference is that NDT or non-destructive testing allows the quality of the weld to be measured, without deforming the part in any way, thus allowing it to be placed back into the process chain if it meets the quality criteria. Ultrasonic and X-Ray testing are two forms of NDT.

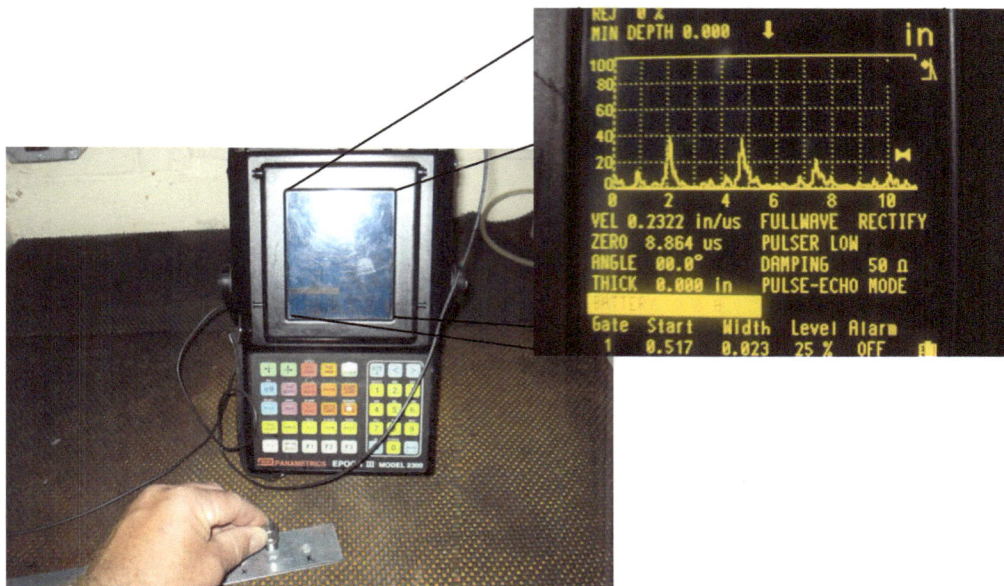

Figure 119 – Ultrasonic weld test system

The transducer for the ultrasonic weld test system is moved over a spot weld, see above coupon sample picture, and a frequency reading is generated. That frequency reading is then compared to the weld frequency quality examples (see figure 120, below) to determine if the weld is acceptable. This process is repeated for each weld on the part.

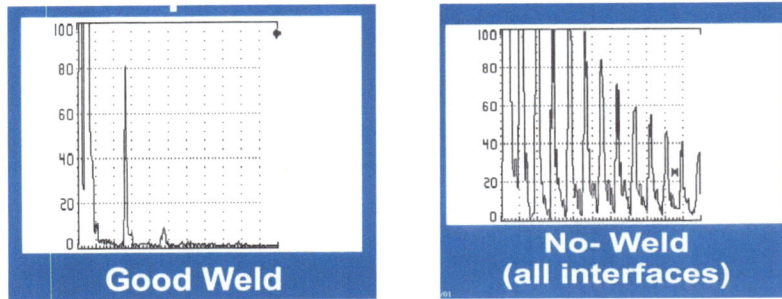

Figure 120 – Ultrasonic weld test results

Destructive testing, on the other hand, is just that. The part needs to be destructed to determine the quality of welds, meaning it becomes scrap after the test is compete. Some forms of destructive weld testing include pull tests, cut and etch testing, and chisel tests.

Below are some pictures of parts before and after "chisel" destructive testing and what the technician is looking for when performing these tests.

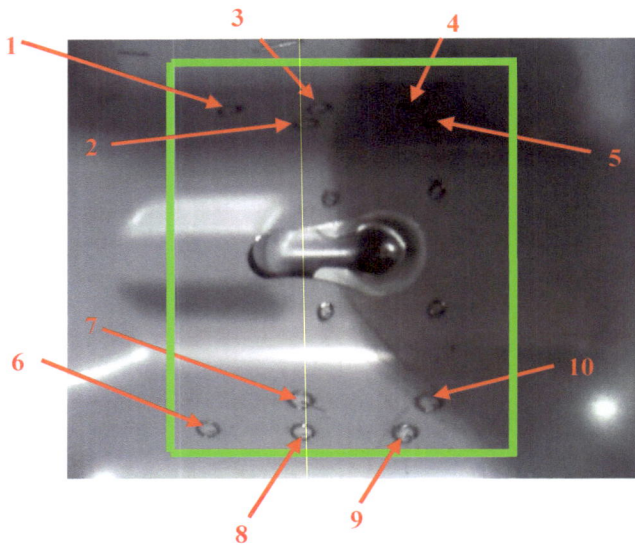

Figure 121 – Before destruct testing

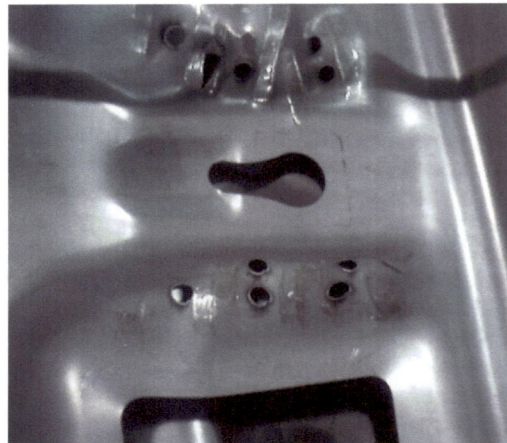

Figure 122 - After destruct testing

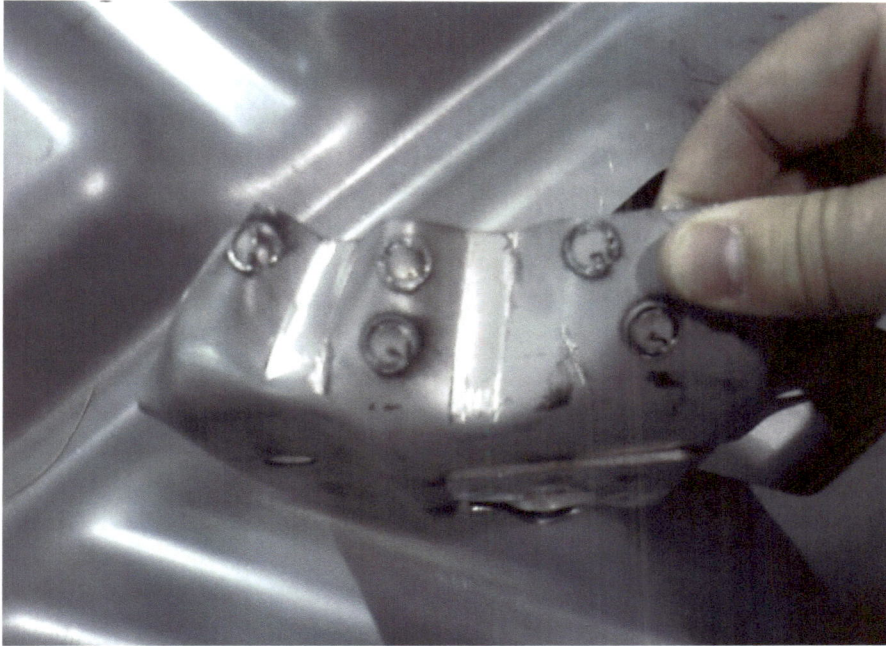

Figure 123 – Underside part (outlined in Green in figure #112)

In figure 121, there are 10 welds that will be destruct tested. The goal is to pull "parent" metal, meaning that when pulling the part off on the underside (see figure 123), metal from the other part will be pulled with it, as seen by the holes in figure 122. These welds are numbered on a check sheet and identified as "pass" or "fail" for each weld. A non-passing destruct test, shown in figures 124 and 125, identify the #2 weld failed as it left parent metal on the main part.

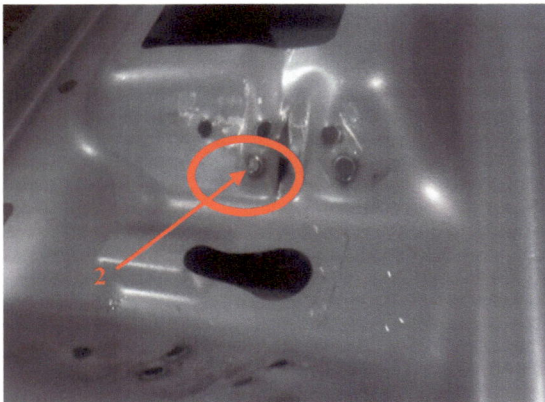

Figure 124

Figure 125

In some cases, depending on the control plan criteria, not all welds need to "pass" this test. The product may be designed with a few extra welds to hold the parts together for all stresses it is designed for, thus allowing a couple to fail and the process still pass overall. This doesn't mean that the process shouldn't be corrected, it should, but it provides the ability to approve parts for shipment.

Another method of destructive testing, cut and etch, shown in figures 126 through 128, are where parts are cut into sections at the mig welds. Acid is then applied to the weld area to "etch" the weld material. Once this is done, the part is either measured electronically through a scope or manually. Below (figure 126) is an example of measurement electronically.

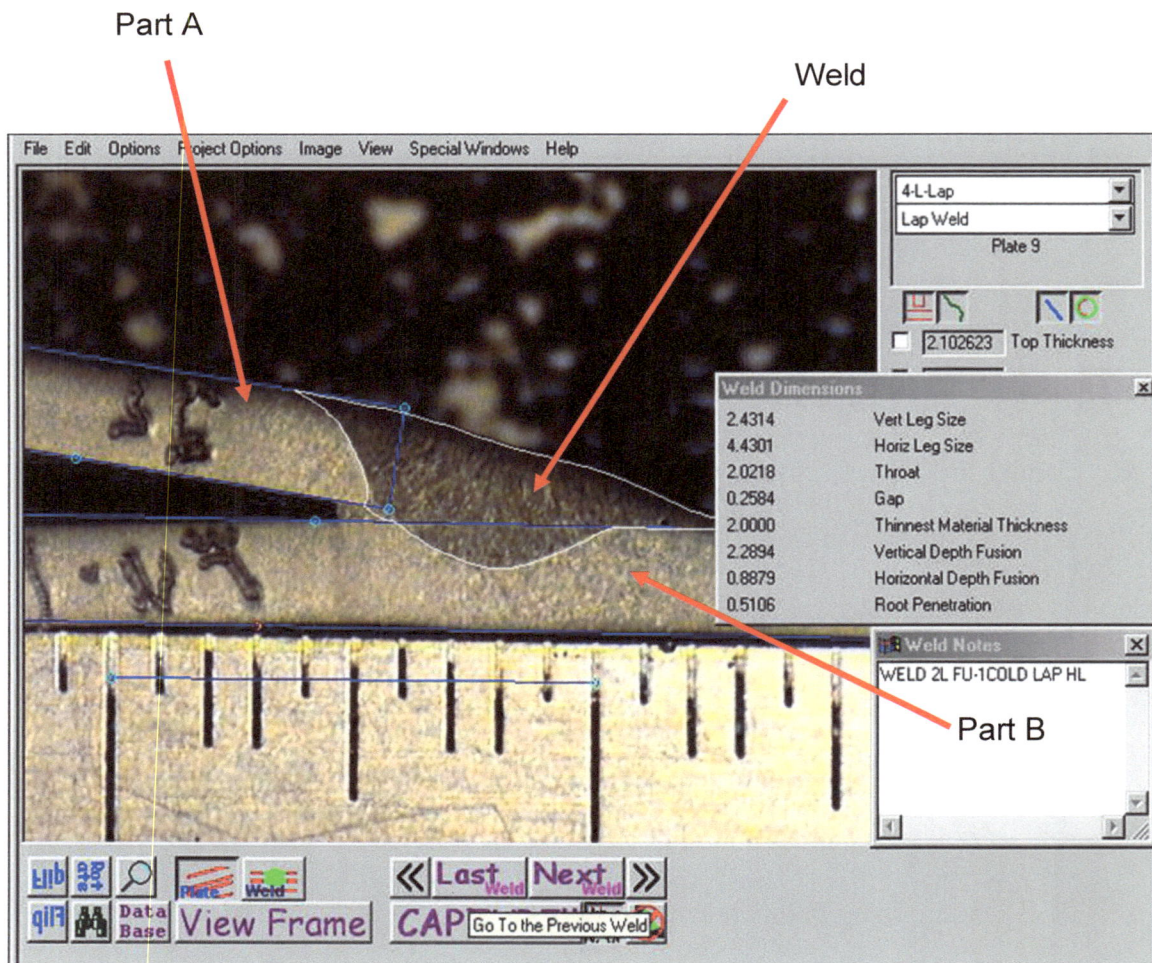

Figure 126

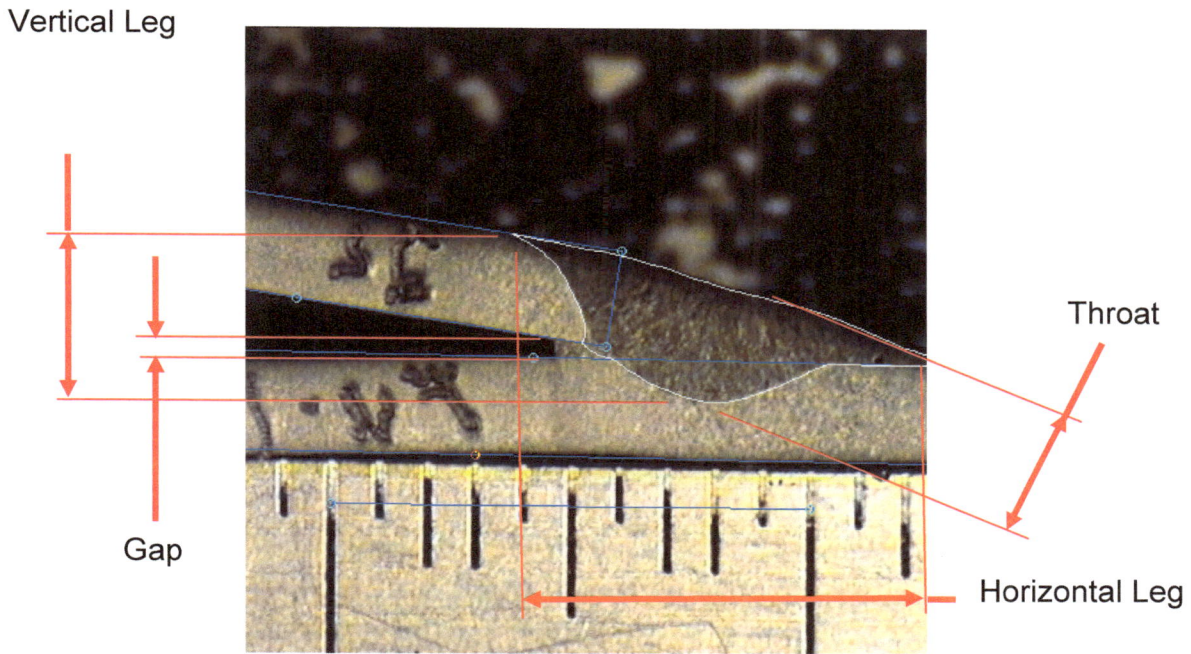

Figure 127

Figure 127, is a brief description of some of the measurements identified in figure 126. The benefit of this particular type of measurement is that the data is stored along with a picture of the part measured, which may be acceptable in place of storing the actual cut parts. These same measurements can also be accomplished using manual calipers, though the accuracy and repeatability of the operator measuring with the calipers may not be as accurate.

Production

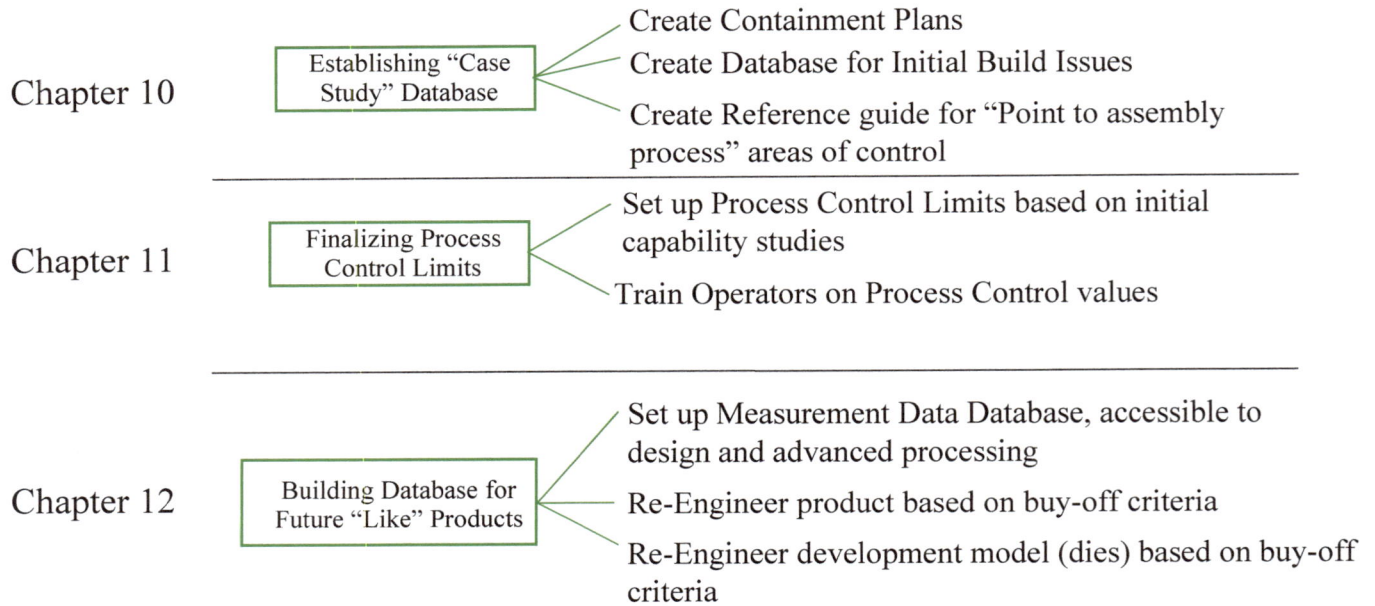

Chapter 10 — Establishing "Case Study" Database
- Create Containment Plans
- Create Database for Initial Build Issues
- Create Reference guide for "Point to assembly process" areas of control

Chapter 11 — Finalizing Process Control Limits
- Set up Process Control Limits based on initial capability studies
- Train Operators on Process Control values

Chapter 12 — Building Database for Future "Like" Products
- Set up Measurement Data Database, accessible to design and advanced processing
- Re-Engineer product based on buy-off criteria
- Re-Engineer development model (dies) based on buy-off criteria

Chapter 10 – Containment Plans

Because a product cannot be considered acceptable by reviewing dimensional data alone, it is important to understand what other means are necessary to ensure that the customer is protected from a defect.

Containment plans are exactly what the name implies. What is the plan to contain a potential issue from leaving? Leaving can be separated into quite a few areas, some examples are:

    a. Leaving one operation to the next
    b. Leaving the manufacturing facility
    c. Leaving the incoming shipping area

Containment plans can range from many items and methods from adding sensors in an operation to adding a dock audit for manually inspecting both incoming and outgoing product.

The best approach to containment depends on specifics for the operation being contained, is it a manual process? Can someone/something review the result in the same automated process? Once some of these questions can be answered, then the containment plan, which will/can coincide with the control plan, be developed. In any case, it is important that the containment plan be considered and developed starting from the process design phase. This way the following items can be considered in the plant strategies:

1. What items need to be contained?
    a. Manually?
    b. Automatically?
2. If manual containment, who will be doing this?
3. Where will the containment area reside?
4. How accessible is this area for material flow, both from the created operation to the next operation?
5. What method is to be used for containment?
    a. Sampled sizes?
    b. 100%?
6. If sample sizes, how will we quarantine parts from the previous acceptable part to the non-acceptable part?
    a. Will the parts be marked via a sequence number, date and time, etc?

Please note: that for each containment action there needs to be a prove-out method. This is some way to verify that the method used for containment will actually work. Also, when designing a manual containment process, it is important to provide enough detail for the operator/inspector to determine if the part is acceptable or not; while not hampering them with too much to do for containment or during the allotted time (i.e. which may be based on the process run at rate).

Manual Containment Plans

Manual containment plans should be created in a way that allows the person reviewing the part the ability to:
1. Distinguish what part they are to look at
2. What areas are concern for containment activities
3. Amount of parts per container/shipment/hour to be reviewed
4. What a good part is versus a bad part is
5. Checklist to validate the action/document results
6. Recommend next course of action, i.e. rework, scrap, etc.

There is a fine line, however, for manual containment plans. These plans need to balance the amount of items being reviewed to the time available and the amount people available to do the task. Manual containment plans need to capture all items that cannot be contained automatically. But, should not be so consuming that they allow an increased possibility of items being missed during the product containment review.

Additionally, a manual containment plan should track:
a.   Sample number(s) (i.e. something that can be traced back to the actual part reviewed)
b.   Method for the reviewer to designate whether the part was acceptable or not
c.   Date and time of the containment inspection
d.   Reviewer's name
e.   Action for parts deemed "not acceptable"

Figure 128 below is an example of a particular part's containment plan. It is the original unfilled document; figure 129 is the corresponding pictorial document that details to the reviewer where the weld nuts exist and the corresponding number sequence.

Weld Nut Clear Thread and Shadowing Containment

Date: _____

Inspector: _____

Shift: _____

Part Serial
Number: _____

Read Procedure For PASSED/FAILED Description

(See Marked Reference Part For Weld Nut Numbering Sequence)

Weld Nut Number	Threaded	Non-Threaded	PASSED	FAILED	Notes
429			☐	☐	_____
430			☐	☐	_____
431			☐	☐	_____
432			☐	☐	_____
433			☐	☐	_____
434			☐	☐	_____
435			☐	☐	_____
436			☐	☐	_____
437			☐	☐	_____
438			☐	☐	_____
439			☐	☐	_____
440			☐	☐	_____
441			☐	☐	_____
442			☐	☐	_____
443			☐	☐	_____
444			☐	☐	_____
445			☐	☐	_____
446			☐	☐	_____
447			☐	☐	_____
448			☐	☐	_____
449			☐	☐	_____
450			☐	☐	_____
451			☐	☐	_____
452			☐	☐	_____
453			☐	☐	_____
454			☐	☐	_____
455			☐	☐	_____
456			☐	☐	_____
457			☐	☐	_____
458			☐	☐	_____
459			☐	☐	_____

Figure 128 – Containment example

Figure 129 – Pictorial of weld nuts for containment list in figure 128

Automatic Containment Plans

Automatic containment plans are methods used automatically in a process to stop a "non-acceptable" product from going to the next operation. This method is usually a 100% inspection meaning all parts in the process go through the containment method. Some examples of this include:

A. Part present switch to verify that a weld nut is present. The presence switch is usually put in the next operation after the geometry set station where the weld nut is originally installed and welded. If the presence fails to detect a weld nut on the part, then a number of things can happen to contain the part from continuing on. For example: from stopping the line until the part is removed and entered into a scrap bin, the part moved automatically to a rework station, or an automatic mechanism tags/paints the part as "non-conforming".

B. Measurement system to read the added feature for location and presence. Using a laser sensor, a measurement can be generated to ensure the part is in the correct location. Also, consequently determines if the part is also present.

C. Scale measurement – the part is placed on a weight scale to determine if the appropriate weight is achieved, which is based on all parts present on the assembly. If acceptable, the part is allowed to continue to the next operation. If deemed unacceptable, the machine can stop and move the part to a rework/scrap bin.

As with any automatic containment process it is important to:
1. Designate what needs to be contained
2. Determine where and how it is to be contained
3. How we intend to verify the containment method works

Item #3 is most important. Why? It is the final method to determine if the containment process used will catch non-acceptable product; not just for the particular failure mode identified, but for all possible potential non-acceptable modes. To accomplish this, we must create an example part for each possible defect, called a RABBET. The rabbets are then used to certify the automatic containment method. This is accomplished usually at line startup each day/shift by placing the rabbet on the automatic containment feature and, in all production modes possible (manual and automatic, if possible) see if the system reacts, as it should.

These rabbet checks should be captured on an inspection sheet each time to ensure they are done and properly working. This can also serve as a containment start point, if a non-acceptable part is found during production, after the last "acceptable" containment process certification was completed.

Overall, while each containment method is acceptable, the manual process relies on the person reviewing the part for approval. In cases of large assembly lines, the only area to review manually is after the line. Thus, a lot of parts can be in process after the offending operation that will have to be quarantined for rework or scrap. Finally, it is an industry accepted standard that a person is only 85% accurate in visual inspection. This means that there is a chance that up to 15% of the production, going through manual containment, may be bad for one reason or another.

Chapter 11 – Setting up Process Control Limits

Once the component has met initial capability requirements, it is time to start setting up control limits, train operators, and trend the process. This will allow you to make adjustments prior to an actual out-of-tolerance issue arises and what tolerance you may need for future like components with similar processes.

Figure 130 below, is a standard SPC (Statistical Process Control) control chart. By using this tool, we can: highlight the tolerance bands, control bands, and track the measured values per the control plan.

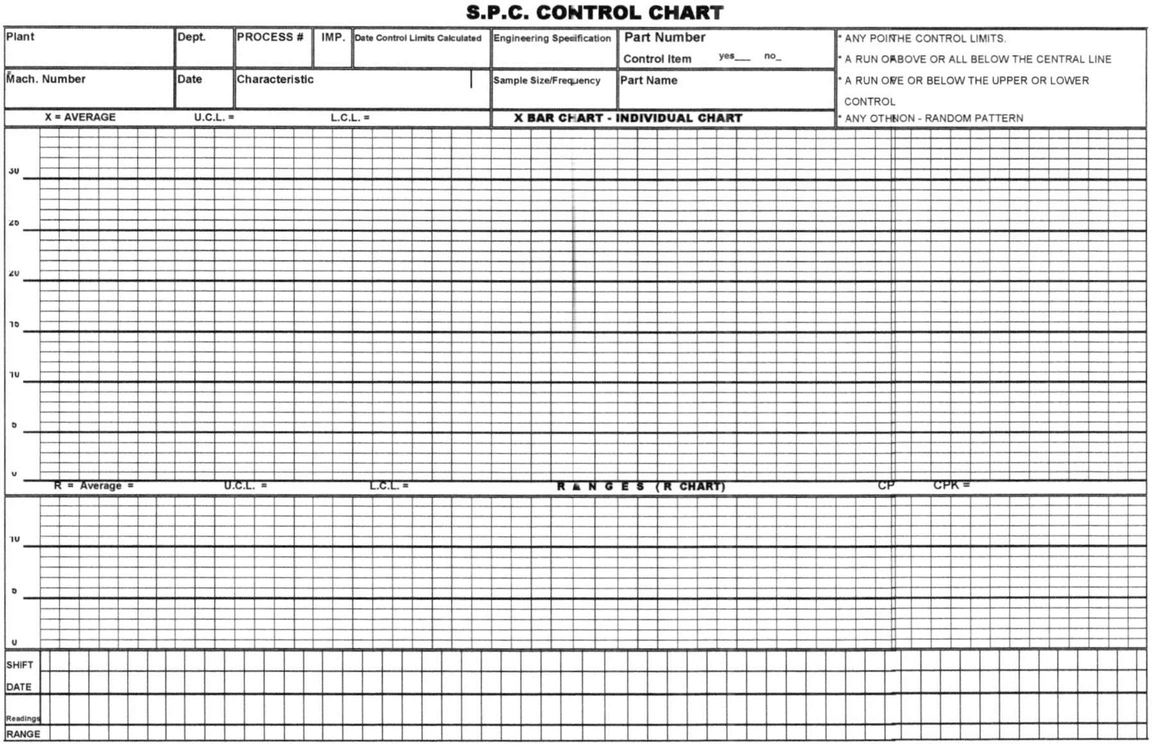

Figure 130

The variable data entered into this chart does not have to be just dimensional measured data. It can also track scrap or rework types in an effort to identify processes and frequencies that may need improvement.

The figure below shows the start of a control chart being used. The red lines represent the change in production shifts. There is also a trendline for all data entered.

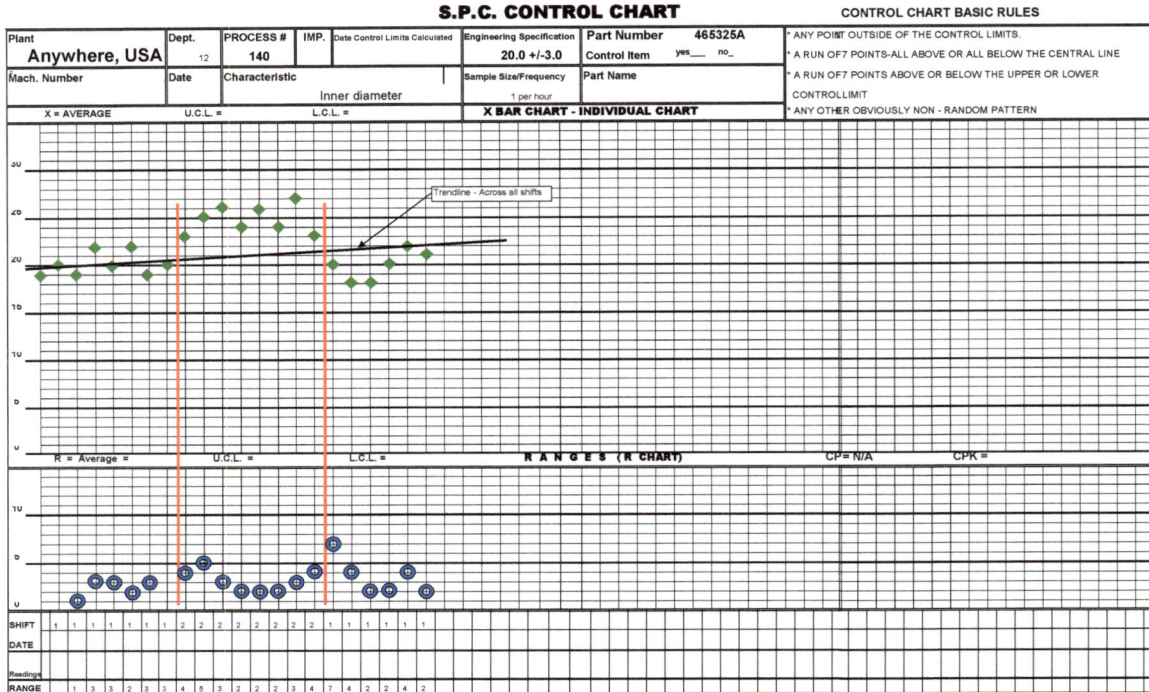

Figure 131

If we look at the measured data alone, we would see from the trend line and from some of the measurements that the process is not in control; but by segregating the data into the appropriate shifts we can easily see that the out-of-control issue is from shift 2. Now we can focus our efforts on shift two operators, maintenance personnel, etc. in order to correct where the change is happening.

While the sample provided above is a basic example, these are the types of identifiers that can be found by properly using and monitoring a control chart.

Another example of a control chart, not used on an hourly basis, but still over time can be used to identify trends in scrap, parts requiring rework, etc.

In figure 132, the operator, per shift, enters specific numbers per hour of scrap or parts requiring rework from their station.

SCHEDULE CONTROL

Date: ______

Shift: ______

Kendallville, IN

Operator Number __________ Pieces: ______

Cell Number 1

Part Number ###

		REWORK				SCRAP									
HOUR	DAILY / PIECES	A1	A2	B1	B2	A1	A2	B1	B2	ORDINAT	Time Down	Time Up	Total	wntime Co	
1st	12	2	4	1	1	1	3	0	0						RWK
2nd	2	0	1	0	0	0	1	0	0						
3rd	3	0	2	0	0	0	1	0	0						
4th	6	1	1	0	0	0	0	0	0						
5th	9	1	2	1	0	1	2	0	0						
6th	9	1	1	1	1	1	3	1	0						
7th	2	1	1	0	0	0	0	0	0						
8th	2	1	1	0	0	0	0	0	0						
Total	45	7	13	3	2	3	10	1	0						

Figure 132

This data is then uploaded into a chart to see any trends over the specific shift, over all shifts, etc. or by specific operation or, whether the operation is performed in more than one spot. Figure 133 is an example of how one shift of data is trended, by fixture, to determine if any trends exist in scrap/rework rates per hour and per fixture (A1, A2, B1 and B2).

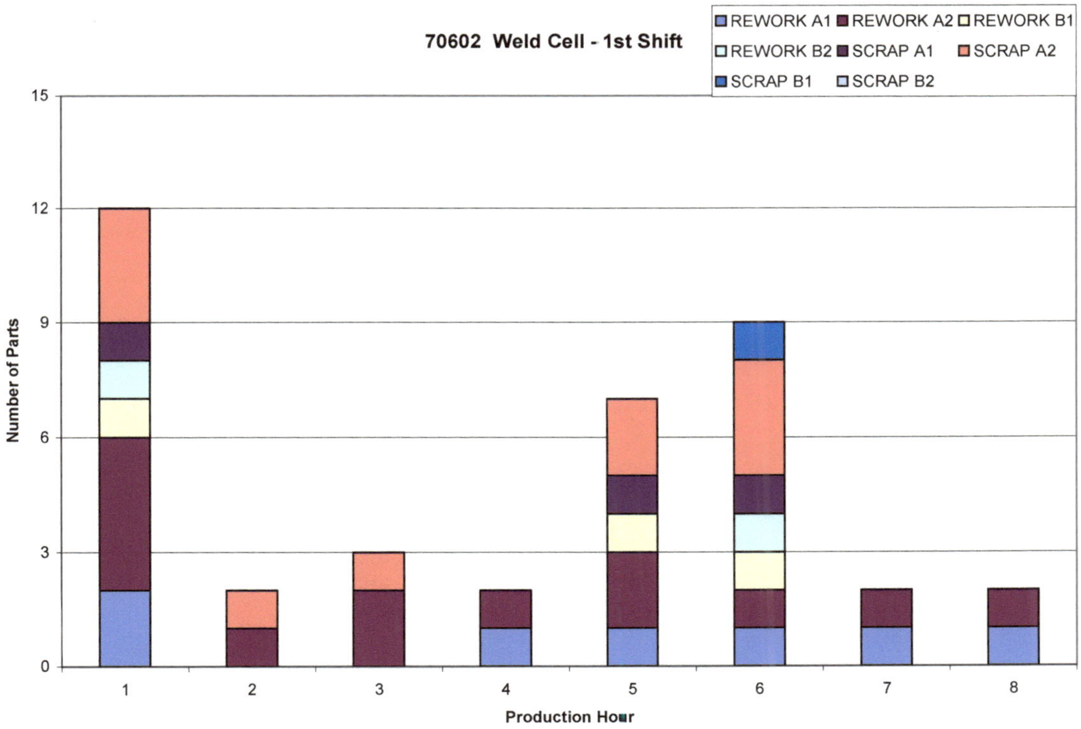

Figure 133

This chart (figure 133) shows that there is "spike" in part issues on the first shift, during the 1st, 5th, and 6th hours. Mostly in scrap and parts requiring rework from fixture A2.

In figure 134, the same data, taken from all shifts, is combined and reviewed by fixture to determine trends per fixture. In this example it appears that fixture "A2" has most of the part issues and needs to be inspected for differences between the other 3 fixtures.

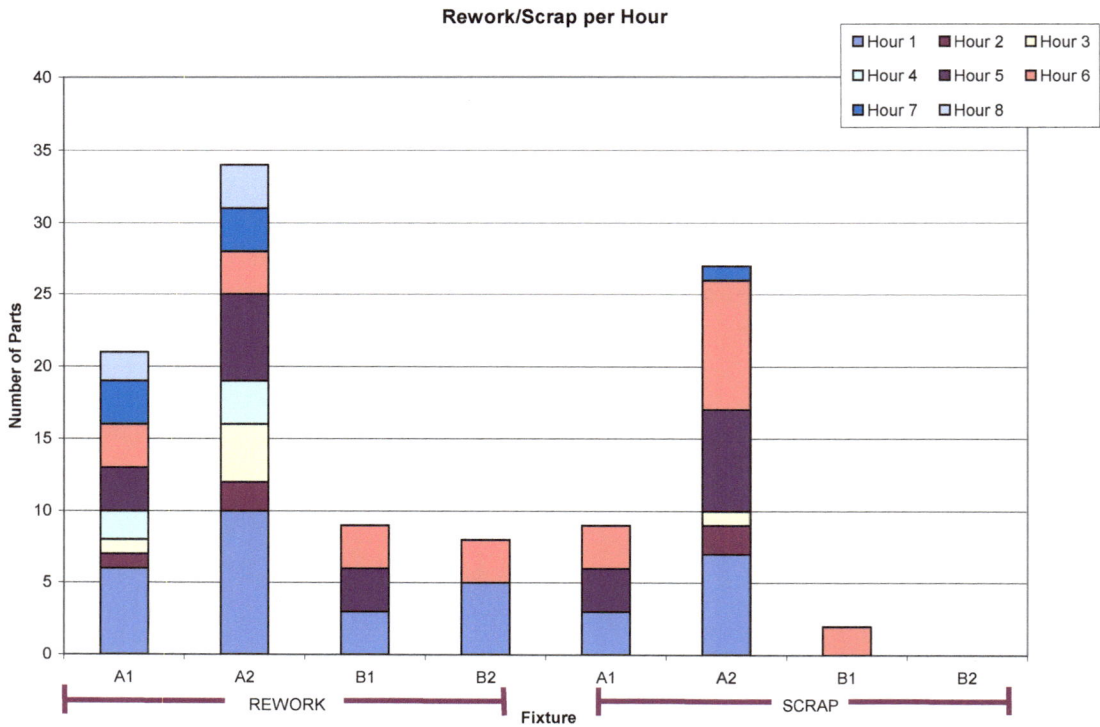

Figure 134

Chapter 12 – Setup Measurement Database

In an effort to reduce "re-creating the wheel", it is recommended that a measurement database be created in a common location so that data can be pushed or pulled into/from it from various organizations within the company. After achieving initial capability at PPAP, it is important to start using ongoing data gathering in a way to better identify tolerances for new product/process designs from an engineering standpoint. Also, to identify and address areas where the data may become close to the control or tolerance limits in the production environment.

For instance, if we create a database for all like components made, say, "crossmembers", we can begin to identify consistent measurement areas from one crossmember to the next. We can then, by adding data accumulated over time, determine long-term capability and then the tolerance needed to achieve it. The results may be different then the original tolerance used at part buyoff due to multiple part setups, material changes, etc.

The database must also include "known" process changes – i.e. tool adjustments, die press changes, etc. so that any mean changes in measurement can be excluded from the overall process tolerance. In the example below (figure 135), we can see over time measurements of a specific point, in this case, in 1-month sample sizes and the overall 3-month sample size. While each month individually may be "capable" the mean shifts experienced from one month to the next due to "known" process changes could make a recommended long-term tolerance larger than it needs to be. But if the month to month mean change is not from a known process change, then it may be considered as part of the overall process variation.

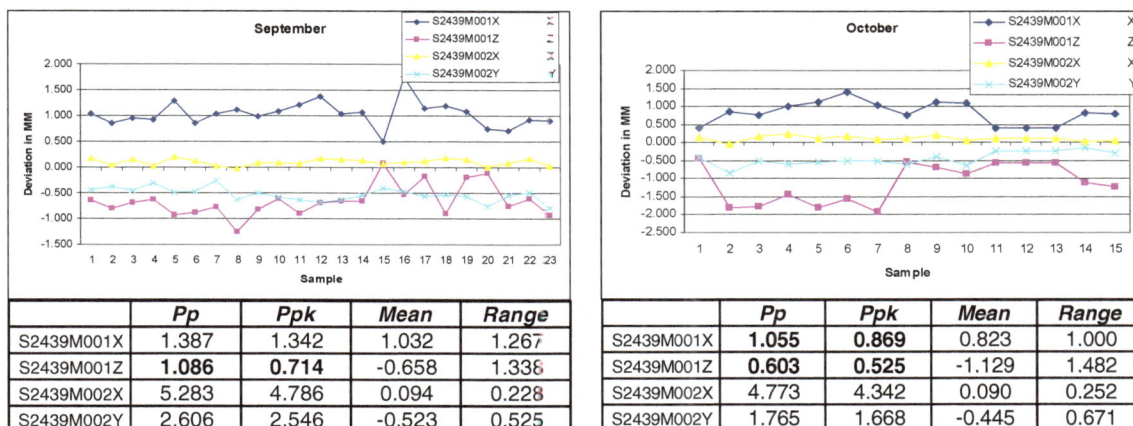

	Pp	Ppk	Mean	Range
S2439M001X	1.387	1.342	1.032	1.267
S2439M001Z	1.086	0.714	-0.658	1.338
S2439M002X	5.283	4.786	0.094	0.228
S2439M002Y	2.606	2.546	-0.523	0.525

	Pp	Ppk	Mean	Range
S2439M001X	1.055	0.869	0.823	1.000
S2439M001Z	0.603	0.525	-1.129	1.482
S2439M002X	4.773	4.342	0.090	0.252
S2439M002Y	1.765	1.668	-0.445	0.671

Figure 135 – 2 months (of 3 months) of data shown individually

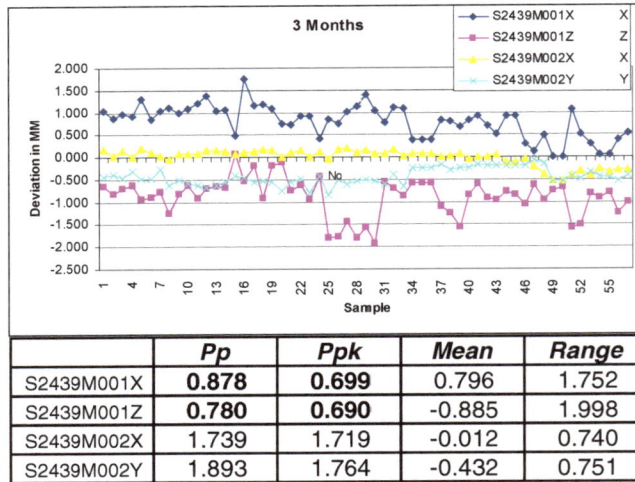

	Pp	Ppk	Mean	Range
S2439M001X	**0.878**	**0.699**	0.796	1.752
S2439M001Z	**0.780**	**0.690**	-0.885	1.998
S2439M002X	1.739	1.719	-0.012	0.740
S2439M002Y	1.893	1.764	-0.432	0.751

Figure 136 – 3 months combined data

Based on the data from figure 136, we have calculated a tolerance to achieve a 1.33Pp based on each month's sample. The overall 3-month sample assuming "known process changes", and the overall 3-month sample assuming "NO known process changes", to illustrate the amount of variation in the tolerance calculation and the importance of understanding what are the underlying factors (known process changes) when evaluating the data for tolerance changes.

	Original Tolerances		September		October		November		Overall	
S2439M001X	2.00	0.00	2.00	0.00	2.10	-0.50	2.00	-0.90	2.50	-0.75
S2439M001Z	0.00	-2.00	0.60	-2.00	1.10	-3.40	0.30	-2.20	0.90	-2.60
S2439M002X	1.00	-1.00	1.00	-1.00	1.00	-1.00	1.00	-1.00	1.00	-1.00
S2439M002Y	0.50	-1.50	0.50	-1.50	0.50	-1.50	0.50	-1.50	0.50	-1.50

Figure 137

In figure 137, point ID S2439M001Z started with a 2.00mm tolerance (total) for September, increased to 4.50mm to meet Ppk requirements for November, but requires a 3.50mm overall tolerance. This is due to various mean shifts in each month, variation in build per month, etc. If the mean shifts are "known" shifts, then the tolerance can be based on the maximum tolerance of any single month, assuming no "known" shifts happened within a given month of data, if not, then the overall tolerance should be pursued.

Additional factors in generating an all-inclusive database may include some of the following examples:
1. Number of dies used to generate the part.
2. Type of die process used (i.e. progressive, transfer, line, etc.).
3. Locating method of the part (i.e. holes vs. form).
4. Which production facility makes the part?
5. Orientation of the part in the production process.
6. Orientation of the part in the measurement process.
7. Production shift part measured on.

Eventually, the ideal method would be to establish standard measurement locations and locators. So, that from part to part, specific features can be measured and compared on like parts across multiple production sites to determine corporate wide overall quality, plus allow for plant-to-plant quality comparisons.

Re-engineering product based on buyoff criteria

In this section we attempt to provide some examples of "closing the loop" when a part is approved or bought off to a different physical condition than that of the CAD model.

Some of the examples we will cover are:
1. Detail part concessions due to the stamping process.
2. Detail part concessions due to the final assembly build approval.

Detail Part concessions due to the stamping process.

In this situation, there may be numerous reasons for a concession. Were surface radii "opened up" to improve material flow? Were surfaces altered to improve process cycle time, with no effect on function and fit items in the next process?

Below are 2 CAD examples – figure 138 being of the original part CAD and figures 139 and 140 showing the actual scan data of the part. The crosshatched areas show where function and fit areas are with respect to other parts in the overall assembly.

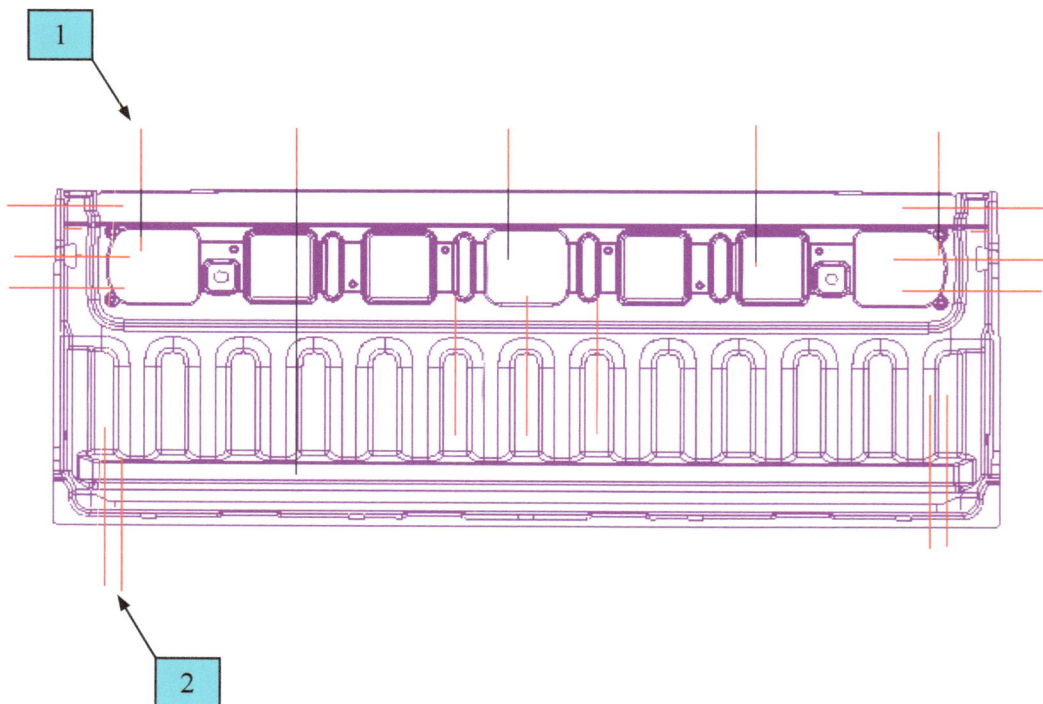

Figure 138

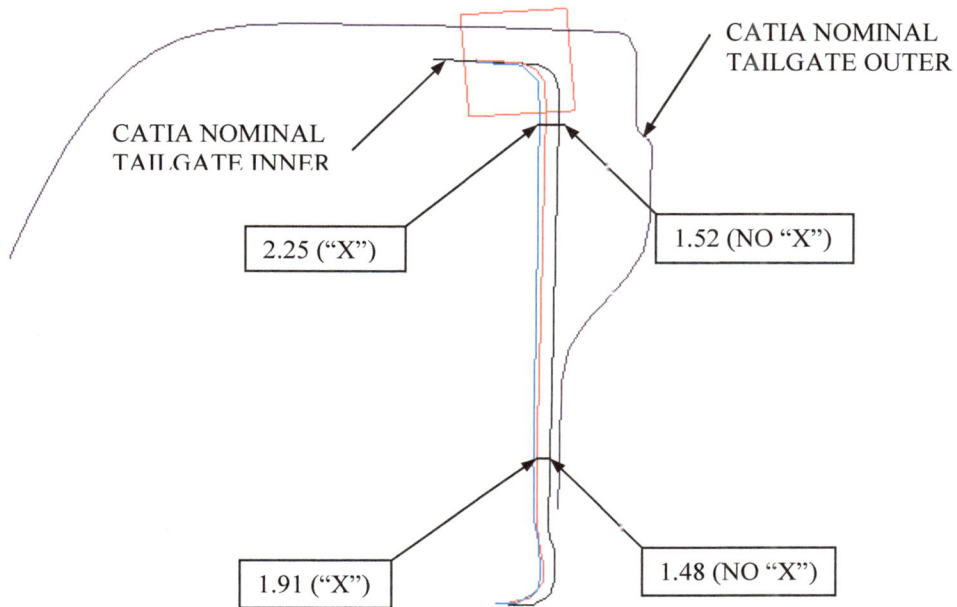

CATIA NOMINAL
TAILGATE OUTER

CATIA NOMINAL
TAILGATE INNER

2.25 ("X")

1.52 (NO "X")

1.91 ("X")

1.48 (NO "X")

LEGEND	
DESCRIPTION	**COLOR**
"X"	BLUE
"NO X"	RED
TAILGATE INNER	BLACK
TAILGATE OUTER	MAGENTA

Figure 139 – Section 1
Note in this example that "X" and "No X" represent (2) draw die cavities in the die process

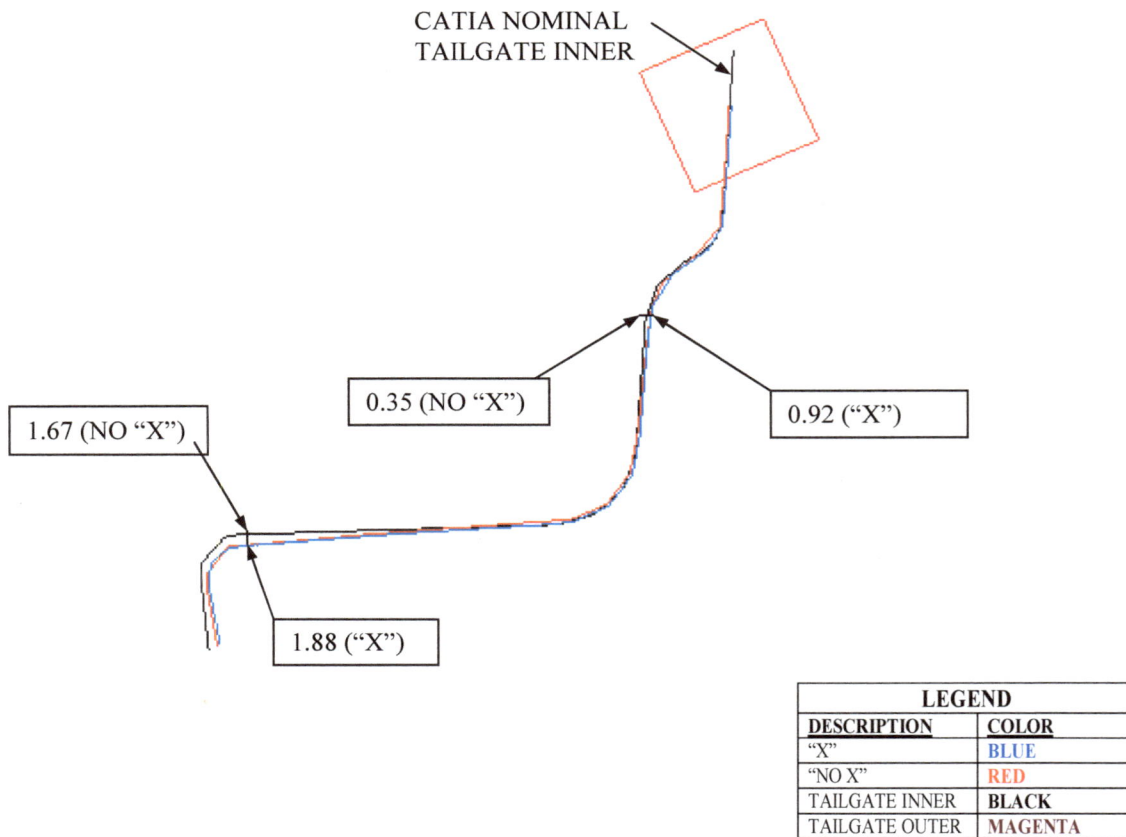

CATIA NOMINAL
TAILGATE INNER

0.35 (NO "X")

1.67 (NO "X")

0.92 ("X")

1.88 ("X")

LEGEND	
DESCRIPTION	COLOR
"X"	BLUE
"NO X"	RED
TAILGATE INNER	BLACK
TAILGATE OUTER	MAGENTA

Figure 140
Note "X" and "No X" represent (2) draw die cavities (same as from figure 139)

In these examples, the new CAD file is generated by the "mean" of the two scanned cavities producing a new nominal in the middle of the measured scans.

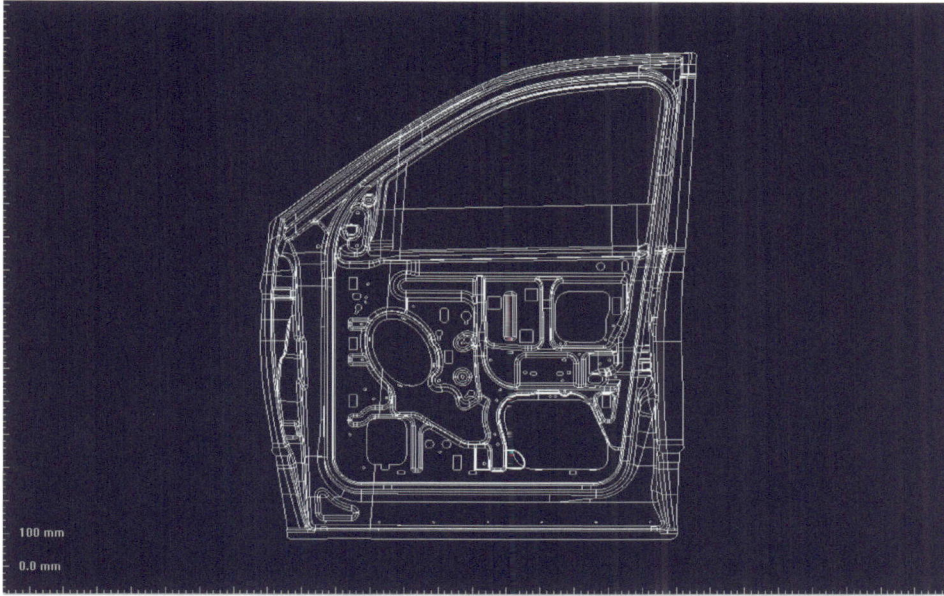

Figure 141 – Product CAD File

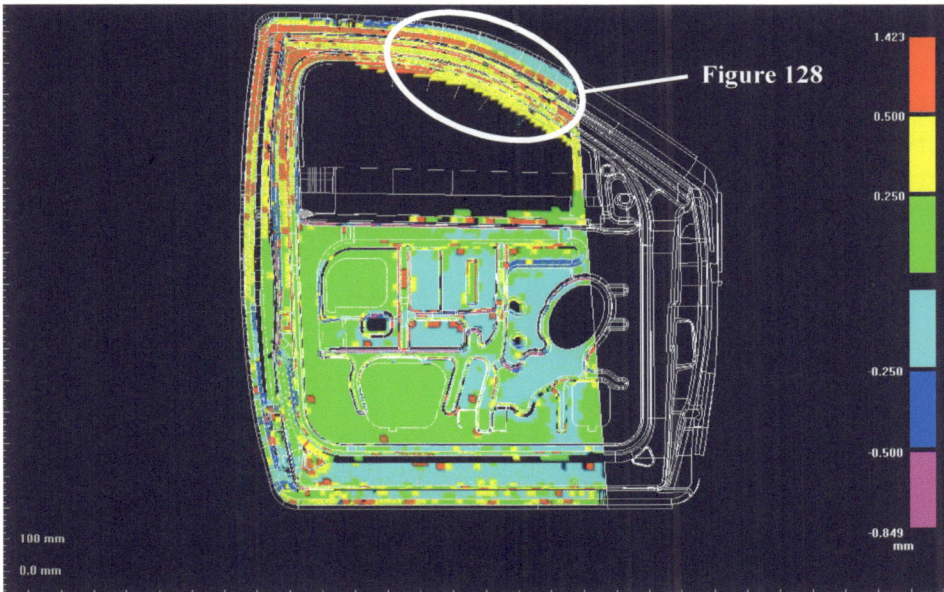

Figure 142 – Scan data comparison to Product CAD file

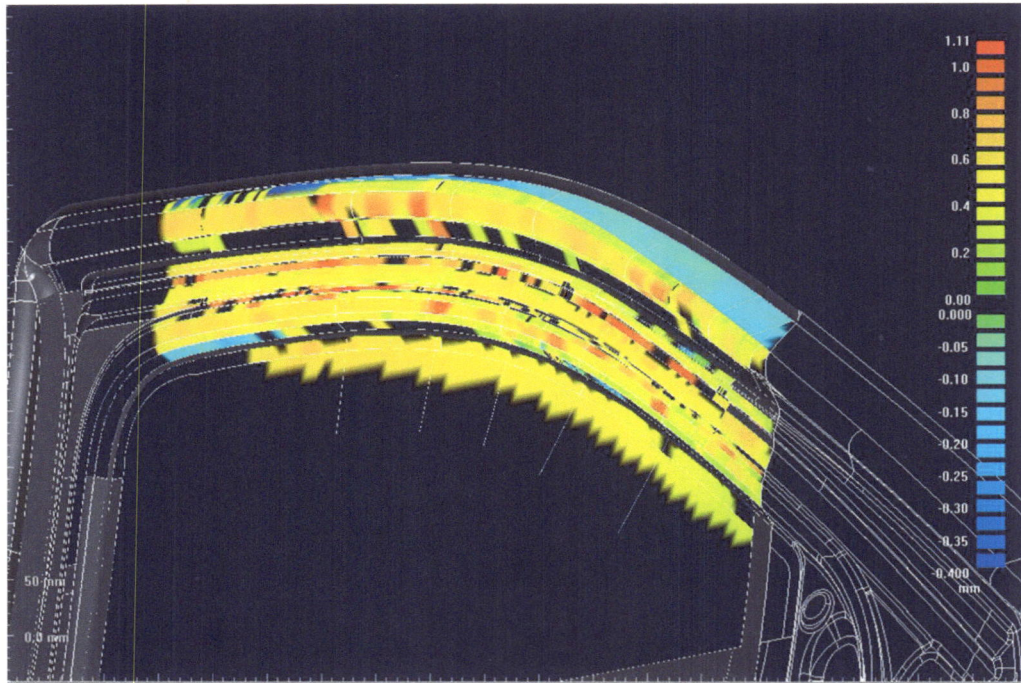

Figure 143 – Closer view of scan data comparison to product CAD file

It is recommended that these changes be followed through some type of engineering change process to ensure that when one group (in this instance, die engineering) requests the "improvement", all other groups (dimensional engineering, assembly engineering, etc.) will be notified and can review and act on what effects this change may cause to their processes.

Re-engineering development model based of the buyoff criteria.

A development model is a CAD surface file used for developing the NC cutter paths for the initial forming process, usually a draw die or crash form die. The development model is created, starting from the original product CAD file, and or adding "compensation" to areas of the part. The objectives are to 1) enable better material flow in the initial forming stage; but will be re-formed in a later stage to get to the product CAD goal, and/or 2) over stretching the product to account for the memory of the material after it is removed from the original forming stage.

Below, in figures 144 through 146 are two examples of these types of compensations from the original product CAD file, to produce to final development model.

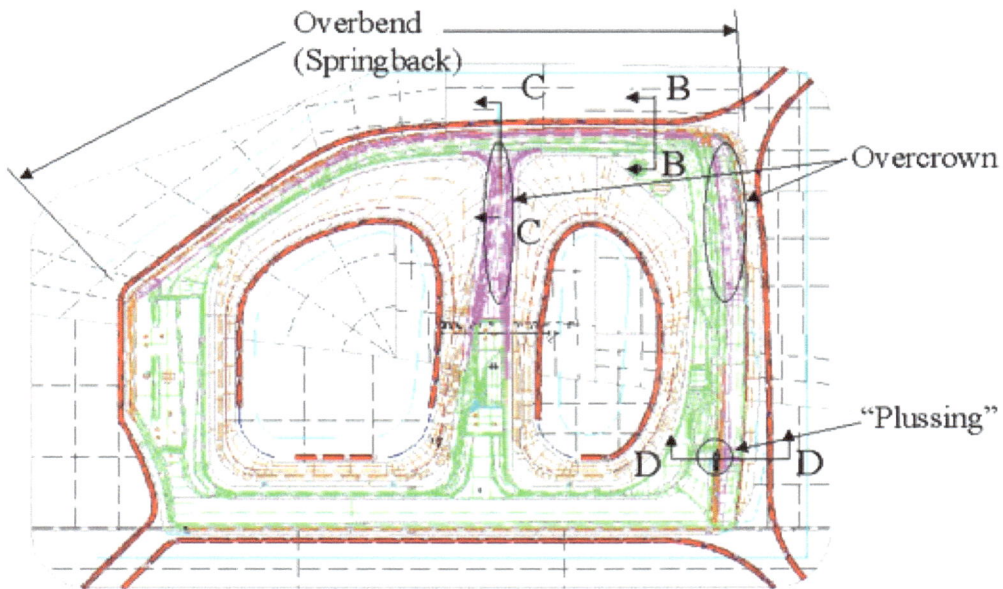

Figure 144 – View of a development model
Details (4) types of compensations used to develop the die forming process.

Plussing (red)

Aperture Outer
Panel (cyan)

Section D-D

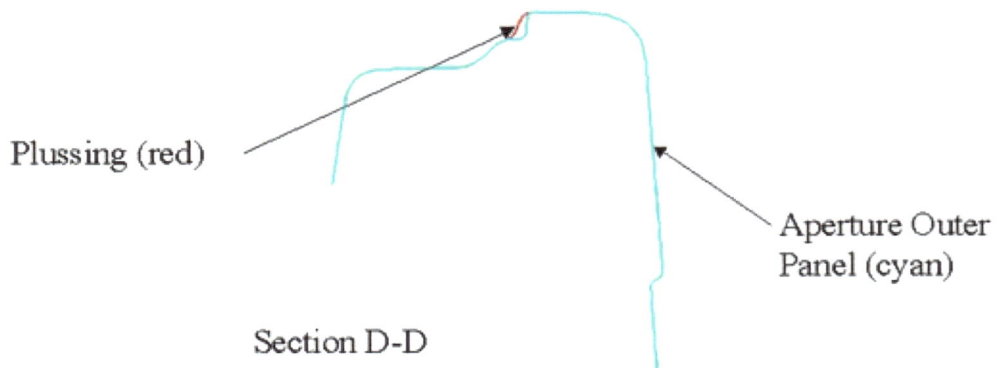

Figure 145 – "Plussing" compensation
Allows for material to flow into the deeper areas of the draw (forming)
operation minimizing thinning and potential for splits.

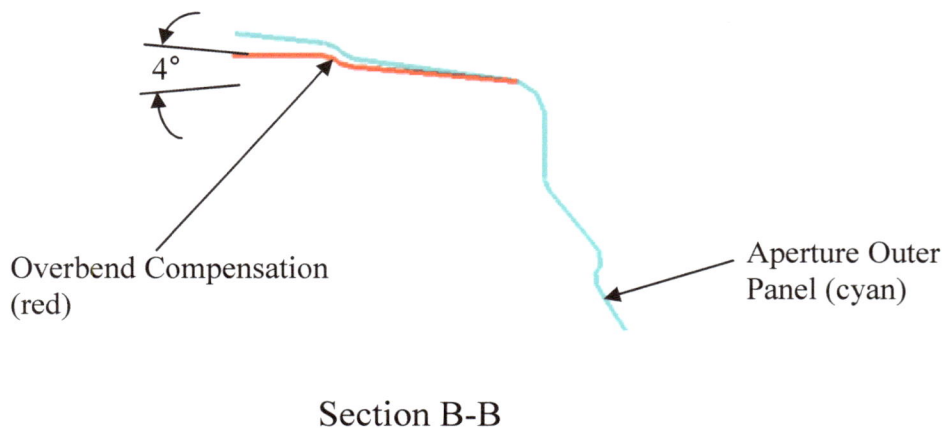

4°

Overbend Compensation
(red)

Aperture Outer
Panel (cyan)

Section B-B

Figure 146 – Overbend compensation
Allows for elastic recovery compensation in a draw die along a face or
wall that is drawn in and is expected to change from the die surface
location. Similar to springback except for a draw die in "local" area.
Final result is that the part is bigger than the die.

All of this is the upfront work done to get an initial part off the die process. If the
part coming off the die process does not agree with the product intent, the dies
are then reworked until the desired output of the product is achieved. It is this
reworking that is important to quantify once the product is approved. To
accomplish this, the die is measured and compared to the original development
model, so that future like components can incorporate the added compensations
into the development model upfront; thus, reducing tryout time. Additionally, this

model can then be used if a second die is to be created for the same part. Or, as part of a maintenance program if the die needs to be re-cut, becoming a "known" origin of producing good product.

Recap

While all the methods/examples used in this book focus on the automotive industry, it is important to understand that the concepts used can be implemented throughout many industries.

The common basis being that proper planning upfront will reduce cost of quality and minimize engineering changes in the build stage. It is just as important to "close the loop" by providing feedback from production builds to original design intent. By closing the loop, the hope is that similar product and process designs of future products will have more realistic expectations from the beginning through tolerances. Additionally, provides an improved understanding of: necessary poka-yokes and containment methods, reduced tool tryout times via improved simulation methods, and reduced overall product timing as a result of these improvements.

Glossary

Terminology	Definition	Chapter	Pages
1D	One-dimensional	2	35
CAD	Computer Aided Design	1	14
CMM	Coordinate Measurement Machines	2	44
Datum Schemes	Locating schemes are developed for two main reasons: to hold the part repeatably for measurement and to hold the part for processing.	2	20
Dimensional objectives	include customer fit and finish expectations on a complete automobile, seal gap or distance on a refrigerator, to the machine finish of a surface on a VCR tape head	1	7
FEM	Finite Element Modeling	2	20,21
Linear tolerance analysis types	1. Worse case, 2. Root sum squared (RSS), 3. Modified root sum squared (MRSS), and 4. Root mean squared (RMS).	2	35
Functional objectives	include door closing efforts, window roll-up speed, etc	1	7
Linear Stack Analysis	two-dimensional only does not require a completed math model as it allows the linear stack to be completed earlier in the production timeline.	2	35, 38
Loop diagram	Starts with the objective and starts at one part looping around to the mating objective part by stepping through each part and process tolerance that is needed to complete the assembly..	2	40
MMC	Maximum material condition	2	25,27
MRSS	Modified Root Sum Squared	2	35
PLP	Principle locating points	2	28
Processing limitations	The limitations surrounding the entire process of making the product itself	1	10
RFS	Regardless of feature size locator	2	25, 26
RMS	Root Mean Squared	2	35
RSS	Root Sum Squared	2	35
Tree diagram	All the "branches" or levels. Each level represents one phase of the build or tear down of the complete objective to the details themselves.	2	41
Point Names	Point identifiers that best describe on what part the point is measured, the type of point, and the direction of measurement	3	45
Point Location	Actual nominal of the measured point, commonly known as the X, Y, and Z-axis location.	3	45
Point Vector	Angle of the surface, which is the normal direction that the CMM probe approaches the point to acquire the most accurate measurement	3	45

| Overbend compensation | Allows for elastic recovery compensation in a draw die along a face or wall that is drawn in and is expected to change from the die surface location. Similar to springback except for a draw die in "local" areas. Final result is that the part is bigger than the die. | 12 | 181 |

Bibliography

The author would like to thank the following people for their contributions:

Michael A Moulton, Thomas Reynolds, Dana Efstate, Fredrick G. Smereka, Timothy Morris, Peter Edmonds, Robert Fegley, Patrick Conway, Daniela Cemalovic and Michael Golembiewski.

References:

Juran Quality Handbook, 4th Edition, 1988.
Measurement Systems Analysis, AIAG Manual, 3rd Edition, March 2002.
Production Part Approval Process (PPAP), AIAG manual, 4th Edition, March 2006.
Statistical Process Control (SPC), AIAG manual, 2nd Edition, July 2005.
Failure Mode Effects Analysis (FMEA), AIAG Manual, 4th Edition, June 2008.
Predicted Versus Actual Compensation in a Stamping Die – SAE Paper, Doc. #2001-01-3108, Published October, 2001
Language of Automotive Body Stamping Dies Defining and Documenting Compensation Within a Stamping Die – SAE Paper, Doc. #2000-01-2713, Published October, 2000
DR Buyoff Training Manual, DaimlerChrysler, 2001.

Special Thanks to:
Geodetics Website - http://geodetic.com/products/products.asp?vstars-m.htm for picture on page 109.
Faro Technologies website - http://www.faro.com/content.aspx?ct=di&content=pro&item=1&subitem=58 for picture on page 109.
Cognitens website - http://www.cognitens.com/44-en/Optigo.aspx for picture on page 109.

www.ingramcontent.com/pod-product-compliance
Lightning Source LLC
Chambersburg PA
CBHW041703210326
41598CB00007B/517